FIRE ALARM HANDBOOK

NICET LEVEL 1 and 2

— ELEMENT REVIEW —

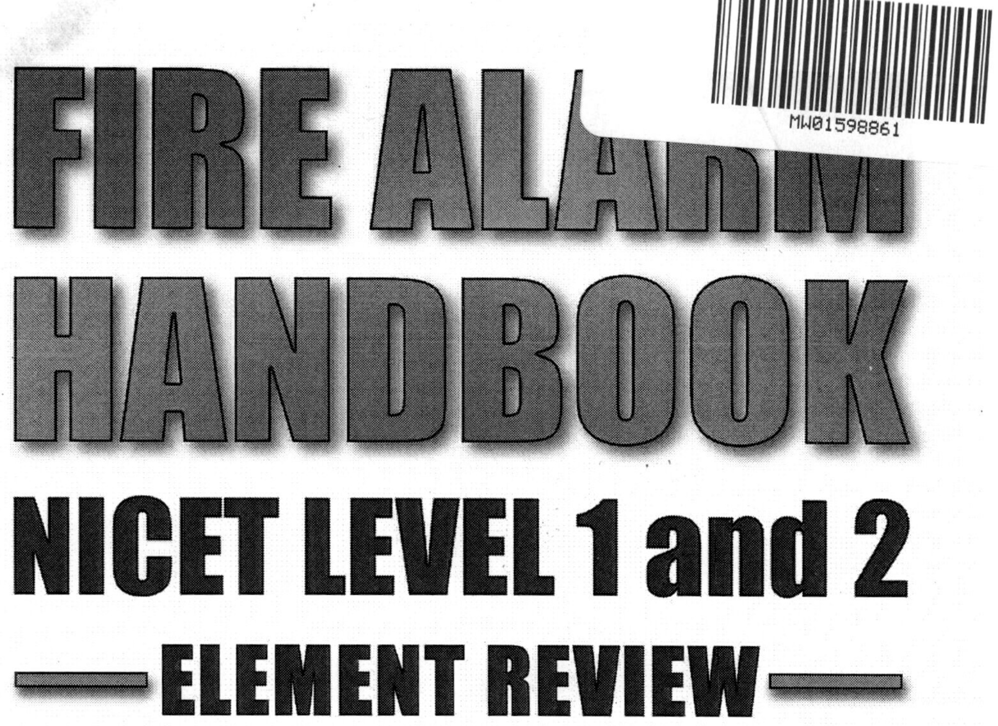

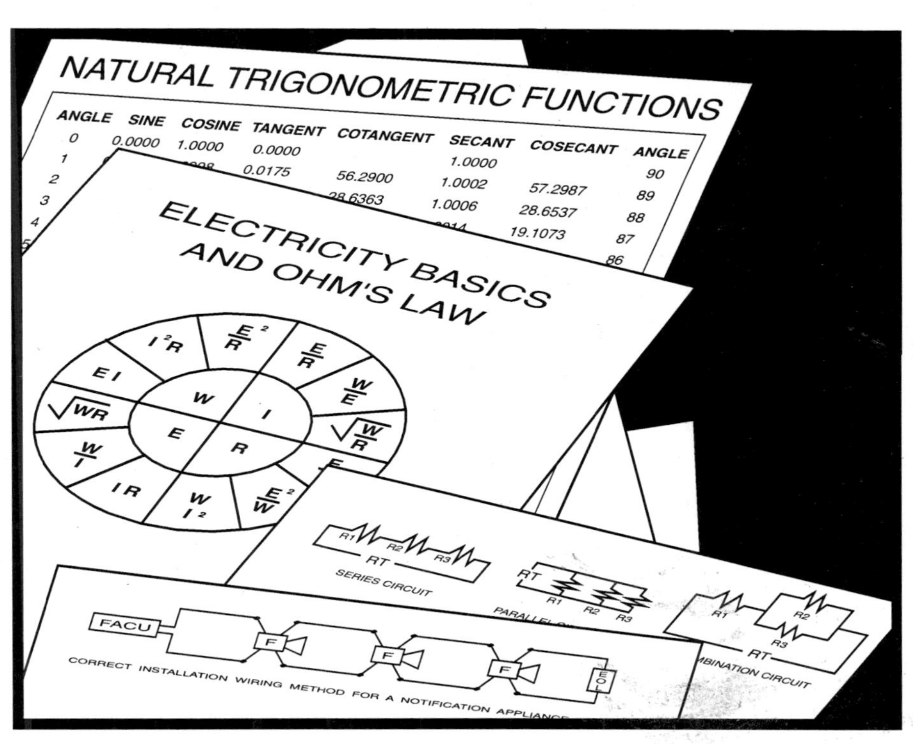

by Grant M. Angell and Michael B. Baker

ISBN 0-9707229-0-7

© Copyright October, 2000
Grant Angell and Michael B. Baker

NOTICE: This Handbook includes references to NEMA, NFPA, SBCCI, and other model-codes publications. We have re-worded these references in an effort to make them easier to understand. Please use the information located in the left column throughout the Handbook to locate the original language used. This Handbook is not to be used as a substitute for any of the publications referenced herein.

TABLE OF CONTENTS

WELCOME! I

FORWARD II

NICET INFO/OHM'S LAW ....TAB 1

LEVEL 1/VOLTAGE DROP FORMULAS...TAB 2

LEVEL 2/NATURAL TRIGONOMETRIC FUCNTIONS....................................TAB 3

GLOSSARY..TAB 4

APPENDIX..TAB 5

NICET INFO

NICET BACKGROUND 1

NICET REQUIREMENTS BY STATE 2

CERTIFICATION BY STATE AS OF 1/1/99 3

9 STEPS TO CERTIFICATION 4

STEP 1: UNDERSTAND THE NICET OPERATIONAL POLICIES 5

STEP 2: REVIEW THE FOUR CERTIFICATION CRITERIA 27

STEP 3: READ THE PROGRAM DETAIL MANUAL 28

STEP 4: REVIEW THE EXAMINATION REQUIREMENT CHART 29

STEP 5: TECHNICIAN PERSONAL RECOMMENDATION FORM 30

STEP 6: COMPLETE THE TECHNICIAN APPLICATION FORM 31

STEP 7: DETERMINE YOUR TEST CENTER LOCATION AND DATE 32

STEP 8: SUBMIT YOUR APPLICATION 33

STEP 9: PRACTICE T'AI CHI TO RELAX AND IMPROVE PATIENCE 34

WHAT TO EXPECT ON EXAM DAY 35

TEST TAKING TACTICS 36

I'VE APPLIED AND TAKEN MY EXAM, NOW WHAT? 37

GREASING THE SKIDS 38

ADDITIONAL RESOURCES 39

FIRE ALARM HANDBOOK
NICET LEVEL 1

LEVEL I ELEMENT REVIEW 41

GENERAL ELEMENTS

ELEMENT	TITLE	PAGE
31001*	BASIC FIRE ALARM SYSTEMS	43
31002*	NFPA STANDARDS	47
31003	BASIC WIRING	49
31004	DEVICES AND COMPONENTS	53
31005*	PERIODIC TESTS	61
31006*	BASIC ELECTRICITY	65
31007	BASIC WORKING DRAWINGS	85
31008	BASIC MATHEMATICS	93
31009	INSTALLATION PRACTICES	99

FIRE ALARM HANDBOOK
NICET LEVEL 1, Continued
SPECIAL ELEMENTS

ELEMENT	TITLE	PAGE
31010	BASIC COMMUNICATION SKILLS	105
31011*	BASIC METRIC UNITS AND CONVERSIONS	107
32001	PLANS, SPECIFICATIONS, AND CONTRACTS	109
32003	BASIC PHYSICAL SCIENCE	113
32004	FIRE WARNING EQUIPMENT FOR DWELLING UNITS	115
32005	BASIC INDIVIDUAL SAFETY	119
32006	FIRST AID PROCEDURES	121

FIRE ALARM HANDBOOK
NICET LEVEL 2

	PAGE
LEVEL II ELEMENT REVIEW	123
LEVEL II ELEMENTS (CONTINUED)	124
LEVEL II ELEMENT (CONTINUED)	125

GENERAL ELEMENTS

ELEMENT	TITLE	PAGE
33001*	FIRE PROTECTION PLANS AND SYMBOLS	127
33002*	BASICS OF SYSTEM LAYOUT	131
33003*	ELECTRICAL INSTALLATION STANDARDS	137
33004*	BASIC FIRE ALARM SIGNALING SYSTEMS	147
33005*	SUPERVISION AND SUPERVISORY SERVICE	151
33006*	DETECTION METHODS	155
33007*	DETECTOR SPACING	163
33008*	POWER SUPPLIES	171
33009	SYSTEM ACCEPTANCE AND PERIODIC TESTS	177
33010	CONSTRUCTION PLANS	183
33011	SPECIFICATIONS AND COST ESTIMATES	187
33012	CONTRACTS	189
33013	BUILDING CODES	191
33014	INSURANCE AUTHORITIES & THEIR REQUIREMENTS	193
33015	GOVERNMENTAL AGENCIES	195
33016	PROTECTIVE PREMISES FIRE ALARM SYSTEMS	197
33017	AUXILIARY FIRE ALARM SYSTEMS	203
33018	SUPERVISING STATION FIRE ALARM SYSTEMS	205
33019	PROPRIETARY SUPERVISING STATION SYSTEMS	209
33020	CENTRAL STATION FIRE ALARM SYSTEMS	213
33021	MANUAL FIRE ALARM SYSTEMS & GUARD'S TOUR SERVICE	217
33022	HEAT SENSING FIRE DETECTORS	221

FIRE ALARM HANDBOOK
NICET LEVEL 2, Continued

GENERAL ELEMENTS

ELEMENT	TITLE	PAGE
33023	SMOKE SENSING FIRE DETECTORS	225
33024	RADIANT ENERGY SENSING FIRE DETECTORS	231
33025	SPRINKLER WATERFLOW AND SUPERVISORY DEVICES	237
33026	ALARM NOTIFICATION APPLIANCES	239
33027	BASICS OF SIGNAL TRANSMISSION	243
33028	BUSINESS COMMUNICATIONS	253
33029	INTERMEDIATE MATHEMATICS	255

SPECIAL ELEMENTS

34001	EMERGENCY VOICE/ALARM COMMUNICATION SYSTEMS	261
34002	SIGNAL PROCESSING	267
34003	SURVEYS FOR FIRE ALARM AND DETECTION SYSTEMS	269
34004	FIRE ALARM SYSTEM MAINTENANCE	277
34005	FIRE ALARM SYSTEM WIRING	281
34006	EMERGENCY EVACUATION SIGNALS	285
34007	COMBINATION SYSTEMS	289

FIRE ALARM SYSTEMS GLOSSARY 291

FIRE ALARM HANDBOOK
APPENDIX

APPENDIX A UL PUBLICATIONS NUMERICAL LISTINGS 323

APPENDIX B UL PUBLICATIONS ALPHAMERIC LISTINGS 339

APPENDIX C NPFA PUBLICATIONS NUMERICAL LISTINGS 355

APPENDIX D NPFA PUBLICATIONS ALPHAMERIC LISTINGS 363

NOTE: General Core Elements noted with *

Notes

A few years ago, a class of apprentice electricians asked me about NICET. Aside from references in NFPA 72, *National Fire Alarm Code,* I knew nothing about it. Out of curiosity I decided to participate in the certification process and report what I discovered back to the class. This step-by-step guide is based upon what I discovered and continue to learn about this important program.

AHJ's throughout the United States recognize the value of NICET certification. Approximately 20 states now require NICET certification of fire alarm system designers, installers, and contractors. Other municipalities such as cities and counties allow either a level III certificate holder or PE (Professional Engineer) to sign fire alarm system design drawings.

NICET does an exemplary job of ensuring an applicant is qualified before granting certification. One common misconception is that the applicant need only pass an exam to become certified. The examination requirement is only one of four criteria used to determine the abilities of the applicant. While it's true that you may attend an exam preparation seminar to prepare for an exam, the three other criteria; work experience, personal recommendation, and supervisor evaluation are seriously considered during the certification process.

In my opinion, the fact that a NICET certificate holder in Maine is given the same consideration as a certificate holder in California is its most valuable feature. As state agencies continue to struggle with electrical licensing reciprocal agreements, the NICET program allows technicians and technologists to rise above it all and pursue job opportunities nationwide.

The fact that you have taken the initiative to find out more about this voluntary certification program says a lot about you and it is my hope that this step-by-step guide will help you through your certification process.

Good luck and have fun!

Michael B. Baker
mbbaker@teleport.com
http://go.to/baker

Notes

Back in 1995 when I first started Limited Energy Resource Center I made a promise to myself. That promise was to try to offer the best training and educational material for the limited energy electrician possible. To that end we started by offering the most comprehensive courses for limited energy licensing from restricted energy technician through journeyman limited energy and including limited energy administrator license test preparation. As new course requirements for NEC update training as will as others came up I followed my firm belief that no one individual has all the answers when it comes to the limited energy field of work. Therefore my next premise went in to effect, find the right person for the job at hand.

After many conversations with Michael Baker it became obvious to me that he was the one to teach the NICET Fire courses. I believed that anyone who worked as a nuclear reactor operator on a sub must have the right drive to develop a fast paced but complete courses in Fire Alarms. Both Mike and myself have dozens of years in the electronic security business and we have known one another for almost 20 years. Mike has always been in the dealership part of electronic security where I have spent a large portion of my time in distribution. With these two varied backgrounds and many nights in front of a computer screen this handbook was produced. It along with over 1000 Powerpoint slides make up the backbone of the NICET Fire Alarm Level 1 and 2 test prep course.

This publication has been completely reformatted using plain English instead of code or legalese language to explain the requirements of the different elements. The side column is used to note where the complete code section can be found in its original language. It was our belief that using normal language would make it easier for the average student to understand the specific requirements of the different elements.

Again, have fun and hopefully we will meet again for Levels 3 and 4.

Grant Angell

GrantLERC@aol.com

Notes

OHM'S LAW & OTHER FORMULAS

W = watts (power)　　　　　　　　　　　　　　　**I** = amps

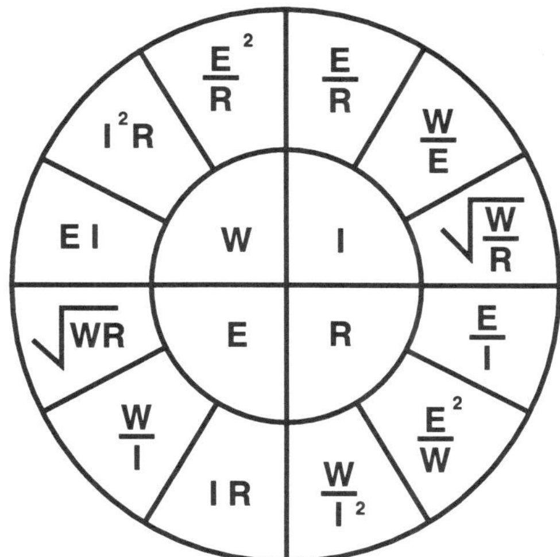

E = volts　　　　　　　　　　　　　　　　**R** = resistance

Ohm's Law Rules

1. When possible try to solve parallels first

2. In a series circuit **voltages** are added, in a parallel circuit **voltages** are constant.

3. In a series **circuit** current is constant, in a parallel circuit **currents** are added.

Kirchhoff's Voltage Law

The sum of the voltage drop is equal to the source voltage

Solving for parallel resistors with a calculator:

1 divided by R1 M+ then 1 divided by R2 M+
1 divided by R3 M+ for as many resistors as needed.
Then to get the answer do: 1 divided by MR = your answer.

The above calculator routine is the same for **capacitors** in parallel.

TEMPERATURE FORMULAS

Celsius > Fahrenheit: $F = C \times 1.8 + 32$　　Fahrenheit > Celsius: $C = \dfrac{F - 32}{1.8}$

LIMITED ENERGY RESOURCE CENTER

CEU and Licensing Prep Classes & Home Study Books
Phone: (503) 657-0135　Fax: (503) 657-2911

NICET Background

- **N**ational **I**nstitute for **C**ertification in **E**ngineering **T**echnologies.

- Founded in 1961.

- A not-for-profit division of NSPE.

- Provides voluntary certification in engineering technology

NICET's mission is to be an independent, internationally recognized evaluator of knowledge and experience; to define and support career paths; and to ensure continued professional development of engineering technicians and technologists.

NICET's vision is to be the leader in recognizing, through certification, engineering technicians and technologists.

The job-task certification program started in 1976 while working on a US Federal Highway Administration contract.

- As of August 2000 over 100,000 certificates were issued in 40 categories.

- As of 4/09/01 7,656 applicants were certified in Fire Alarm Systems.

- Only 67 certificates in Oregon and 144 in Washington as of August 2000.

NICET provides nationally applicable voluntary certification programs covering several broad engineering technology fields and a number of specialized sub-fields.

The value of NICET certification

- Increased professional stature.

- Increased opportunities for advancement.

- Quick entry into job assignments that would otherwise require extensive on-the-job training.

- Opportunities for job assignments where there are regulations or contract provisions that require NICET certification.

NICET Requirements by State

AK   Level II to install, test or maintain. Level III or PE to sign design drawings.

AL   City of Montgomery requires level III/IV or FPE to sign design drawings.

AZ   Gilbert and Scottsdale require a level II to sign design drawings.

CO   City of Colorado Springs requires level III or PE to sign design drawings.
        Cities of Westminster and Aspen require level II to sign design drawings.
        Denver allows any level to maintain in lieu of Electrical Signal license.

DE   Level III or passage of level III work elements required to take State exam.

LA   Fire Alarm Contractor is required to employ a level III.

MA   Level II or higher, or Class DC, or Class D license required.

MI   Fire Alarm Specialty Technician License requires level II.

MT   With manufacturers' certification you may sell, service, and install those equipment
        lines. Level II allows you to work with all manufacturers.

NC   Level II for Fire Alarm Technician and level III for Superintendent.

NE   Level II certification is accepted as an alternative to taking the Nebraska State Fire
        Marshal's exam for Fire Alarm Inspector's license.

NJ   Atlantic City requires level II to inspect, maintain, or repair.

OK   Unlimited Fire Alarm Technician requires level II or NBFAA level 2B.

OH   State allows level III/IV for certification as designer.

OR   The city of Beaverton requires level III or PE to sign design drawings.

PA   Philadelphia will require level II by 1/1/2000 and Level III By 1/1/2002.

SC   Fire Alarm Contractor is required to employ a level III.

TX   Level II for installer and level III for Fire Alarm Planning Superintendent.

WA   The City of Camas requires a level III or PE to sign design drawings.

WV   Requires NICET certification of at least one employee per dealer.

Fire Alarm Certification by State as of 4/9/2001

	Level						Level				
	I	II	III	IV			I	II	III	IV	
**	5	7	2	9	23	MT	11	38	2	3	54
AK	59	151	14	5	229	NC	45	116	31	12	204
AL	15	46	14	5	80	ND	3	8	8	3	22
AR	8	32	13	3	56	NE	9	47	5	3	64
AZ	43	85	27	9	164	NH	8	19	4	5	36
CA	158	279	57	35	529	NJ	39	142	26	16	223
CO	36	73	17	8	134	NM	5	14	1	1	21
CT	14	70	20	1	105	NV	22	42	7	1	72
DE	2	31	14	3	50	NY	71	336	44	23	474
FL	194	258	61	40	553	OH	55	145	48	20	268
GA	57	107	25	17	206	OK	21	66	8	7	102
HI	5	12	0	1	18	OR	34	42	10	6	92
IA	9	19	2	4	30	PA	64	184	65	21	334
ID	9	22	5	3	39	PR	0	2	0	0	2
IL	62	196	30	25	313	RI	2	12	5	1	20
IN	27	75	13	9	124	SC	10	47	13	7	77
KS	16	58	12	8	94	SD	2	9	1	0	12
KY	14	34	24	3	75	TN	28	59	18	6	111
LA	51	107	106	22	286	TX	109	302	158	51	620
MA	28	99	22	21	170	UT	14	20	7	3	44
MD	61	103	43	23	230	VA	55	74	26	24	179
ME	2	12	3	2	19	VT	1	7	0	1	9
MI	54	412	28	19	513	WA	39	100	15	14	168
MN	42	113	13	9	177	WI	21	65	17	6	109
MO	33	98	12	8	151	WV	27	12	2	1	42
MS	2	11	4	2	19	WY	2	4	2	2	10

Level I	Level II	Level III	Level IV
1,700	4,322	1,103	531

Total - 7,656

9 steps to certification

Step 1: Understand the NICET operational policies.
These policies are the NICET organizational rules. While it isn't necessary to memorize these, you should familiarize yourself with the policies and pay particular attention to policy #30, Continuing Professional Development.

Step 2: Review the four certification criteria.
Successfully completing the examination is only one requirement.

Step 3: Read the Program Detail Manual.
A Program Detail Manual exists for each sub-field and includes valuable information. Use the element descriptions and list of references to prepare for the examination and then use the tally sheet to track of your progress.

Step 4: Review the examination requirement chart
Carefully plan for your examination to avoid some common pitfalls such as neglecting to test in a core-element.

Step 5: Complete the Technician Personal Recommendation Form.
Seriously consider whom you may recruit to be your recommender.

Step 6: Complete the Technician Application Form.
Use this form to request an examination, submit additional work experience information, change your address of record, etc.

Step 7: Determine your test center location and date.
A blank table is included for the test centers in Oregon and Washington.

Step 8: Submit your application.
If you are testing, don't forget to include the fee.

Step 9: Practice Tai Chi to relax and improve patience.
Good things come to those who wait.

Step 1: Understand the NICET Operational Policies

The first listing of NICET Operational Policies was published in 1992. In 1995, some of the policies were revised and some additional policies were added. Below are the policies, as they existed in 1995, except for Policy #4 and Policy #30. Policy #4 is revised effective March 1998 and Policy #30 was revised effective October 1997.

Please be advised that all policies are currently under review and changes will be forthcoming.

POLICY 1. COMPLAINTS AGAINST APPLICANTS AND CERTIFICANTS

Introduction

NICET certification is granted when the applicant for certification has demonstrated, through examination and submission of qualifications, the knowledge and skills required to properly function in a capacity relevant to the certification. **NICET** does not investigate the day-by-day performance of **NICET** certificants because it is the certificant's employer who has the duty to monitor and evaluate actual performance. **NICET** certification is a credential, which is acquired after an evaluation of demonstrated pertinent skills and knowledge -- **NICET** does not, in any manner, warrant that certificants will function properly in a capacity relevant to the certification.

Complaints Accepted and Possible Actions

Three types of complaints will be accepted and they may be registered by any party, including NICET itself. The three types of complaints are:

Type I. Complaints may be accepted pertaining to an applicant's or certificant's qualification for a particular certification. These complaints shall pertain to examination misconduct; false, incomplete, or misleading application information; or false or misleading personal recommendation. Possible actions may include temporary suspension or permanent revocation of certification(s).

Type II. Complaints may be accepted pertaining to violations of the **NICET Code of Responsibility** by a certificant. Possible actions may include reprimanding the certificant.

Type III. Complaints may be accepted pertaining to a certificant's improper execution of the technical practices, which are an integral part of the certification. Possible actions may include temporary suspension of certification(s) or reprimanding the certificant.

Processing of Complaints procedures:

1. Complaint will be reviewed by the NICET staff to ascertain if the criteria for acceptance of a complaint are met. The acceptance criteria are as follows:

    A. Complainant must have personal knowledge of the alleged violation or misbehavior or must be in a position to supply relevant and reliable documentation.

B. Complainant must demonstrate by documentation and factual evidence that the complaint involves an issue or issues directly related to the criteria set forth in the preceding section entitled "Complaints Accepted and Possible Actions." Matters of a personal nature or matters not related to the criteria set forth will not be considered.

C. The complaint will not be processed if the NICET records show the named person is no longer in the NICET Registry.

D. Complainant must file the NICET prescribed complaint form no later than two months from the date complainant first became aware of the alleged violation, and no later than six months from the date of the alleged violation.

2. Accepted complaints will be processed by the NICET staff for the purpose of attempting to resolve the complaint informally.

3. If resolution by the NICET staff is not possible, a NICET Hearing Panel will be established to review the complainant's statements and, if necessary, formally solicit answers to specific questions from the applicant or certificant. The Hearing Panel's ruling will then be conveyed in writing to the applicant or certificant and the complainant. The Hearing Panel shall be constituted as follows:

A. One current NICET Board of Governors member, one former NICET Board of Governors member, and one certificant (SET or CT, as appropriate). The appointments shall be made by the NICET Chairman and shall be made with the objective of having two of the three panel members familiar with the applicant or certificant's certification technical area.

1) No Hearing Panel member shall be appointed to serve for a case in which he or she has a conflict of interest or under any circumstances in which the member cannot render impartial and unbiased judgment.

2) The Hearing Panel may, at its discretion, retain a special consultant familiar with the applicant's or certificant's area of expertise if it believes that the special consultant will assist it in its deliberations.

4. If the Hearing Panel's ruling is unsatisfactory to the applicant or certificant, a Formal Hearing may be requested before the Hearing Panel. The request for a Formal Hearing must be received in writing by NICET no later than 30 days after the issuance of the Hearing Panel's ruling. The Formal Hearing will be scheduled between 60 and 90 days after NICET's receipt of the request for the Formal Hearing.

A. NICET will do its best to ensure that the Formal Hearing will be held at a time and location convenient to the applicant or certificant.

1) In the event the Formal Hearing is postponed, it shall be re-scheduled and take place within six months of the issuance of the Hearing Panel's original ruling.

B. The applicant or certificant will be informed of the rules for the Formal Hearing, which will include the following:

1) A written description of the scope of the hearing including a clear statement of the allegations to be heard will be given to the applicant or certificant.

2) The applicant or certificant will have the right to have legal counsel and to call witnesses.

3) The proceeding will be tape recorded by NICET.

4) The applicant or certificant will have the right to submit written testimony in lieu of appearing at the Formal Hearing provided such intent is made known thirty days prior to the scheduled hearing date.

5) If written testimony is submitted, the hearing will proceed on the scheduled date whether or not the applicant or certificant is present.

C. If the applicant or certificant does not appear and written testimony is not provided, no Formal Hearing will be conducted and the original Hearing Panel's ruling will be final and non-appealable.

5. The Formal Hearing will proceed whether or not the applicant or certificant is present if written testimony is submitted.

6. If the Formal Hearing results in a decision which is unsatisfactory to the applicant or certificant, a final appeal to the **NICET Board of Governors** may be requested provided a written request is submitted to the **NICET General Manager** not later than 20 days after the Hearing Panel issues a written decision based on the Formal Hearing. This final appeal shall be scheduled at a regular Board of Governors meeting, which is convened not later than ten months after the date the Formal Hearing was conducted.

7. All decisions of the NICET Board of Governors are final and non-appealable.

Public Notice of Actions

Final actions will be reported in the NICET Newsletter.

Definitions

1. TEMPORARY SUSPENSION is an action that prevents testing activity, certification eligibility evaluations, and certification decisions in any certification area. Additionally, the individual will not be listed in any certification directory published during the suspension period. The length of a suspension period is influenced by the situation, but the minimum period is six months.

2. PERMANENT REVOCATION is an action which terminates the individual's association with NICET. Prior testing records, certification records, etc. are purged and future association with NICET is blocked.

POLICY 2. HANDLING OF CERTIFICATION PROCESS IRREGULARITIES

Introduction

NICET seeks to be fair and consistent to all applicants during the process that leads to a certification decision. However, situations may arise in which either NICET or the applicant may dispute some aspect of the certification process.

The purpose of this document is to describe NICET's procedures for handling such disputes. The terms "temporary suspension" and "permanent revocation" as used in this policy are the same as defined in Policy #1.

A. Examination Misconduct

1. Prior Knowledge

A. If NICET has reasonable cause to believe that an examinee has obtained knowledge of any of the contents of his or her scheduled examination or any other examination prior to or during an examination sitting, NICET will cancel the examination results of the examinee and any examinees who have also obtained knowledge of any of the contents of the examinations and will not permit these persons to sit for any NICET exam for a period of two to ten years.

2. Copying Questions

A. If NICET has reasonable cause to believe that an examinee intentionally compiled or copied information from an examination during its administration for the purpose of passing this information on to future examinees, NICET will cancel the examination results and will not permit the examinee or those known to have received the information to sit for any NICET exam for a period of two to ten years.

B. False, Incomplete, or Misleading Information on Application

If NICET's review of the information on a certification application form reveals that false, incomplete, or misleading information which was intentionally included on the application contributed to or would contribute to an improper certification decision, NICET may take the following action:

1. Delay the certification decision.
2. Temporary suspension of certification(s) held.
3. Permanent revocation of certification(s) held.

C. Improper Work Element Verification

If NICET determines that any verification was obtained from a non-qualified verifier or was given for tasks not actually performed, NICET's action against the applicant or certificant may be one of the following:

1. Delay the certification decision.
2. Temporary suspension of certification(s) held.
3. Permanent revocation of certification(s) held.

In addition, NICET may take action against the verifier, which may include one of the following:
1. Written reprimand to the verifier
2. Permanent loss of privilege of serving as a verifier.
3. If NICET certified, temporary suspension or permanent revocation of certification(s).

D. Improper Personal Recommendations

1. If NICET determines that the submitted recommendation form was prepared by a non-qualified recommender, NICET's action against the applicant or certificant may be one of the following:

    a. Delay the certification decision.
    b. Temporary suspension of certification(s) held.
    c. Permanent revocation of certification(s) held.

2. In addition, NICET may take action against the recommender which may include one of the following:

    a. Written reprimand to the recommender.
    b. Permanent loss of privilege of serving as a recommender.
    c. NICET certified, temporary suspension or permanent revocation of certification(s).
    d. All decisions of the appointed panel are final and non-appealable.

E. Work Experience Evaluation

If the applicant registers dissatisfaction with NICET's evaluation of the submitted work experience, the work history of record will be reviewed in the following sequence of events:

1. Reevaluation by the NICET Deputy General Manager with decision in writing to applicant.

2. If resolution does not result, evaluation by the NICET General Manager with decision in writing to applicant.

3. If resolution does not result, an evaluation will be made by a panel appointed by the NICET Chairman. The panel shall consist of three persons with experience in the sought after area of certification. When possible, the panel shall have one current or former NICET Board of Governor member, one individual who is a manager or supervisor from an environment similar to that in which the applicant works, and one certificant (SET or CT as appropriate) working in an environment similar to that in which the applicant works.

4. All decisions of the Panel are final and non-appealable.

POLICY 3. RELEASE OF CONFIDENTIAL APPLICANT/CERTIFICANT INFORMATION

NICET will make every reasonable effort to prevent an individual posing as someone else from obtaining privileged information.

1. If the requestor is seeking information on an individual other than himself or herself, only the following information will be released:

    A. Certification status information will be limited to:

    > 1) certification is in Active, Retired, Delinquent, or Expired Status.
    > 2) certification grade (level).
    > 3) certification field/sub-field.
    > 4) certification number.

    B. Applicant status information will be limited to confirming that the individual is, in fact, an applicant.

2. If the caller is requesting information on him/herand the NICET staff member can establish that the caller is actually the certificant or applicant in question, the staff member can confirm or clarify information on record.

POLICY 4. CERTIFICATION STATUS

If your NICET certificate indicates that it is valid through July 1, 2000 (OR LATER), Policy #4 does not apply to you. You must comply with the requirements of Policy #30.

Policy #4 was revised effective March 1998. Paragraph #6 was removed from this policy and now exists separately as Policy #31.

1. **Active Certification Status**. A certification is maintained in **ACTIVE STATUS** by payment of the annual renewal fee. If a certificant has been awarded more than one certification, payment of the annual renewal fee is required for <u>each certification</u> to maintain Active Status in each. Failure to pay the required annual renewal fee for Active Status certifications by March 31 of the renewal year will result in a status change to **DELINQUENT STATUS**. If a partial payment is made on the annual renewal fee, the certifications paid for will retain their Active Status and those not paid for will be put into Delinquent Status.

2. **Retired Certification Status**. Retired Status is available for persons who have retired from all paid employment. All certificates held must be moved into Retired Status and a flat fee is charged regardless of the number or type of certification(s) held. There is no reinstatement to the Retired Status if the fee payment is not made by December 31

of the billing year. A person who wishes to return to Active Status from a current Retired Status must pay the Reinstatement Fee plus the full renewal fee for Active Status.

3. **Delinquent Certification Status**. A certification can remain in Delinquent Status for 3 consecutive calendar years. If a Delinquent Status certification is not **REINSTATED** prior to the end of the third year, the certification will be changed to **EXPIRED STATUS**. A certification in Delinquent Status cannot be upgraded to a higher level. Additionally, a certification in Delinquent Status cannot gain or provide crossover credit.

4. **Expired Certification Status**. When a certification goes into Expired Status, all work element credit associated with that certification is lost. The certification can be regained only by reapplying as a new applicant and meeting the current criteria for the certification. Successful applicants following this course of action will be reissued their original certification number. It is possible to have one certification expire, but retain other certifications through continuing payment of the annual fee for those particular certifications.

5. **Certification Reinstatement**. A Delinquent Certification can be returned to Active Status by paying the annual renewal fee due for the current billing year and paying the reinstatement fee. These payments must be accepted by **NICET** on or prior to December 31 of the third year before Expired Status is imposed on January 1 of the 4th year.

POLICY 5. RETIRED CERTIFICATION STATUS

Incorporated into Policy #4.

POLICY 6. CERTIFICATION AT LEVEL IV

All candidates for certification at Level IV will be required to meet the criteria for all lower levels of certification. There will be no direct entry at Level IV without meeting the exam requirements for level II and III.

POLICY 7. DISTRIBUTION OF NICET MAILING LISTS

NICET does not sell or indiscriminately distribute mailing lists of NICET applicants and/or certificants. This general restriction includes requests from persons seeking potential employees. However, under certain limited conditions, NICET will provide names and addresses of definable groups of applicants and/or certificants for one time use to bonafide organizations upon determination by the NICET Board of Governors that such use will be of direct benefit to the targeted group.

POLICY 8. EXAMINATION QUESTION SECURITY

Access to the NICET certification examination questions will be limited to the following persons under strict security restrictions:

1. NICET staff
2. NICET program advisory committee members
3. NICET question writers and reviewers
4. Bonafide instructors preparing training materials
5. Organizations preparing a specialized test.

The conditions under which access will be permitted and the persons qualifying for access are solely and exclusively determined by NICET.

POLICY 9. STATUS OF EXAMINATION REQUIREMENT COMPLETION

NICET will NOT advise certification candidates of the number or type of work elements still required to complete an exam requirement. If a certification candidate neglects to keep a personal tally of work elements or misplaces score reports, he or she may purchase copies of previously issued score reports or a Personal Transcript, which shows all work elements credited to the test history (by test, crossover and external credit).

POLICY 10. PROGRAM DETAIL MANUALS AND GENERAL INFORMATION BOOKLET

Upon publication of a new edition of a program detail manual or the general information booklet, all previous editions are considered obsolete. New editions may introduce rules and procedures that are effective immediately or on a prescribed date.

POLICY 11. CERTIFICATION CRITERIA EFFECTIVE DATE

An evaluation to determine whether a candidate complies with all the applicable certification requirements will only be made after an examination requirement has been satisfied.

If an examination requirement has been satisfied, but one or more of the other criteria have not been satisfied, then a Conditional Decision Letter regarding certification will be mailed to the candidate. This conditional decision will be honored for up to one year from the date of issuance. After the one-year period, the candidate must comply with any new criteria that may have come into effect in the interim.

POLICY 12. NON-REFUNDABLE FEES

Application, renewal, and other fees must be pre-paid and are not refundable. Fees may be changed by NICET without prior notice.

POLICY 13. CERTIFICATION ELIGIBILITY – CITIZENSHIP

NICET certification is open to all individuals residing in the United States and its territories and to all U.S. citizens living abroad who meet the certification requirements.

POLICY 14. CERTIFICATION ELIGIBILITY - WORK EXPERIENCE

Certification will be awarded only to those individuals who have appropriate engineering technician or engineering technologist level work experience.

POLICY 15. EXAM APPLICATION AND PAYMENT OF EXAM FEE

Exam applications must be postmarked no later than the date listed on the "Test Centers and Exam Dates" schedule. Postmarked deadlines will be strictly observed. Applications postmarked after the published date will be held for the next testing date at the center selected, unless a second choice with a later deadline date is indicated. The exam fee payment in full, must accompany the exam application. Applications for testing which are not accompanied by full payment will not be processed until full payment is received. Processing of applications, which lack critical information, will not be completed until unless the deficiencies are corrected in accordance with the time frame provided in the notifying letter.

POLICY 16. FOREIGN WORK EXPERIENCE CREDIT

The NICET work experience requirement must be met by a preponderance of the work experience being acquired while residing in the U.S. or its territories. Applicability of the work experience obtained in a foreign country will be determined on a case-by-case basis to establish the extent to which U.S. standards and work practices were followed but the rule-of-thumb is that a maximum of two years of experience credit will be awarded.

POLICY 17. PERSONAL RECOMMENDATION FORM VALIDITY

Personal recommendation forms will remain valid for twelve months after the date that the recommender signs the form. Post-dated forms remain valid for twelve months after the date NICET receives the form.

POLICY 18. EXAMINATION BY SPECIAL ARRANGEMENT

NICET allows the administration of certification exams at locations other than its published test sites only if NICET does not have a published test site within 75 miles of the examinee(s) residence(s) and the following conditions are met:

1. The requestor of the special arrangement can offer a suitable location for administering the exam(s).
2. NICET can obtain the services of a test proctor who will not be perceived as having a connection to the examinees.
3. The requestor submits the "Special Arrangement Form" to the NICET office at least 75 days prior to the desired test date.
4. If the requestor is sponsoring more than one examinee, the applications for the examinees must be sent to NICET in one package at least 50 days prior to the desired test date.
5. The requestor remits to NICET in advance of the testing date a special arrangement fee that covers a proctor honorarium and any other necessary charges to administer the examination(s).

The requestor must accept the responsibility of:

1. Delivering the examinee test confirmation letter(s) in a timely manner;
2. Notifying each examinee of the testing location and reporting time;
3. Developing with the proctor a plan for notifying examinees for situations which require changing the test date.
4. Negotiating directly with examinees if deferrals and/or re-schedulings are needed for a test date. NICET will not negotiate with the examinees, only with the requestor.

Any applicant who lives within 75 miles of a NICET test center is not eligible for a special testing arrangement.

POLICY 19. AMERICANS WITH DISABILITIES ACT

Applicants with a disability as defined in Title III of the Americans With Disabilities Act who may be placed at a disadvantage when taking a NICET certification examination must advise NICET in writing of their needs by including a letter and/or other appropriate information along with their application to test.

POLICY 20. RETESTING OF FAILED WORK ELEMENTS OR PART A OR B EXAMS

Failed work elements and Part A/Part B exams can only be rescheduled for examination after a period of 140 days has elapsed. Retesting a failed work element or Part A/B will be limited to three attempts to achieve a passing score. After the third failure, NICET will require written notification that the applicant has obtained training pertinent to the work element or the Part A or B exam topics before further testing will be permitted.

POLICY 21. TESTING OF CROSSOVER WORK ELEMENTS

Work elements, which are listed as crossovers are considered to be the same or very nearly the same work element. Thus, examinees that test a work element, which is a crossover to a previously tested work element, must use the same guidelines as are provided in Policy #20 above.

POLICY 22. RELEASE OF EXAMINATION SCORE REPORT

The certification Examination Score Report is released only to the examinee. NICET will release certification examination scores to employers and others only if the examinee submits NICET's release form to authorize release of the scores.

POLICY 23. POSTING OF CROSSOVER CREDIT TO TEST HISTORY

Crossover credit to and from work elements in Active Status certifications is recorded as of the date a test is scored and stored electronically in the examinee's test history. A crossover relationship that was terminated prior to the test date or that was established after the test date will not be recorded. Crossover credit will not be assigned to or from work elements if the certification is in Delinquent or Expired Status.

POLICY 24. EXAM RESCHEDULING

Examinations, which are scheduled for administration at published NICET test sessions may be **DEFERRED** or **POSTPONED**, as follows:

Conditions for deferral (no additional fee required) are:

1. The examinee determines as unacceptable the "make up" test date, which NICET scheduled after a test, is canceled due to unacceptable conditions at the test site, weather conditions, or proctor unavailability.

2. The examinee determines as unacceptable a future test date, which NICET assigns, because the examinee did not indicate a second choice on the application and the first choice was at capacity when the application was processed.

3. The examinee properly notifies NICET in writing <u>no later than 19 days before the scheduled test date</u> that he/she cannot sit for that exam.

When these conditions for deferral are met, no additional fee is required to reschedule an exam. A deferral, however, has a limited time span. The deferred exam must be taken within 8 months of the initially scheduled test date.

An exam, which is deferred by the examinee (#3 above), cannot be deferred a second time by the examinee.

Conditions for postponement (rescheduling fee required) are:

1. The examinee initiates a post-ponement for whatever personal reason applies (medical emergency, family emergency, work emergency, etc.) by properly notifying NICET during the period from 18 days before the test date to 5 business days after the test date. There is no limit to the number of times an exam can be post-poned, but the rescheduling fee must be paid each time the exam is post-poned.

Conditions which require another full exam fee and a new application to reschedule an exam are:

1. An applicant who calls or writes to CANCEL a scheduled exam.
2. An applicant who is a "NO-SHOW" on the scheduled test date and who does not contact NICET within five (5) business days after the test date. In both these cases, the exam application is deleted.

POLICY 25. REACTIVATION OF TESTING RECORDS

All records of testing, regardless of field or sub-field, will be deactivated when a period of three (3) years passes from the date of the last test taken without any NICET testing activity and the applicant has not qualified for any NICET certification. Testing records will be reactivated only upon payment of the reactivation fee.

POLICY 26. DELETION OF TESTING RECORDS

All test records for an individual certification area (work element or Part A/B) will be purged from the live database after five years if no further testing is done in that certification area and the if the individual is not certified in that certification area.

If the applicant has active certifications or is actively testing in other certification areas, those certification areas will not be affected.

POLICY 27. COPYING OF COMPLETED FORMS

NICET will not accept faxed or duplicated copies of already-filled-out forms. The forms themselves may be duplicated prior to being filled out if you need extra copies, but we will only accept the filled-out form if the information is original (typed or handwritten).

POLICY 28. SEALS AND STAMPS

The use of any seal or stamp conveying the NICET name or mark on engineering documents or drawings prepared or checked by a NICET certificant is illegal.

POLICY 29. SEATING AT REQUESTED TEST CENTER

Seats at test sessions are given out on a first-come, first served basis according to when the application is received. Due to limitations on spacing at some test centers, NICET cannot guarantee seating of all applicants requesting a particular test session, even if the applicant complies with the postmark deadline.

POLICY 30. CONTINUING PROFESSIONAL DEVELOPMENT

1. Recertification through Continuing Professional Development (CPD) is required to continue each and every certification in Active Status beyond the expiration date of the certification.

2. The certification period is three years (36 months).

    A. Certificants with multiple certificates will have the same expiration date for all certificates.

    B. When a certification is upgraded to a higher level, the certification period associated with the lower level certification is continued without change.

3. Each certification loses its Active Status on the last day of the expiration month printed on the certificate.

4. Recertification is achieved by accumulating a prescribed number of continuing professional development points (see Exhibit I) within the certification period through professional development activities directly related to the following:

    A. Being an active practitioner in the certification practice area.

    B. Acquiring additional education pertinent to the certification practice area.

    C. Participating in activities, which advances or broadens the body of knowledge for the certification practice area.

    D. Actively is seeking to upgrade certification(s) held and/or actively seeking initial or upgrade certification in related practice areas.

    E. Successfully completing a special (certification maintenance) written examination.

5. Activities undertaken to acquire CPD points for a particular certification may be used (in whole or part) to acquire CPD points for other certifications if the activities are nationally recognized as common to the certification practice areas in question.

6. Recertification requires the recertification application form and the recertification fee payment to be received by NICET **no later than 80 days** prior to the expiration of the certification.

7. Information provided on the recertification application forms will be processed as follows:

A. Information given will be reviewed for understandability, applicability, and comprehensiveness. Clarifications may be requested.

B. Selected recertification applications will be audited and those certificants will be required to substantiate the claims made on the application forms by way of reproductions of documents that validate the claims and written statements from reputable persons verifying particular activities.

1) certificants selected for an audit shall be identified by a combination of a fixed and random sampling plan that insures that certificants are identified from each state and that within each state, certificants holding the more popular certifications are identified.

2) audits shall be conducted **prior** to the end of the certification period so that decision and appeal rulings are made before the next certification period begins.

3) certificants selected for audit will be notified on or before 65 days prior to their certification expiration date.

4) certificants selected for audit must return to NICET all the requested documentation on or before 50 days prior to their certification expiration date.

8. The fee for recertification must be paid in advance.

A. The recertification fee is not refundable in whole or in part if recertification is denied.

B. Certificants with more than one active certification shall pay the base recertification fee and an incremental fee for each additional certification for which application for recertification has been made.

9. CPD activities must be accomplished before the certification expires.

A. The only exception allowed for a certificant is:

1) additional time required to acquire CPD points because certificant had to allocate substantial time to a critical need such as military duty, jury duty, or convalescence that arose with essentially no warning in the last 12 months of the certification period and prevented accumulation of the remaining necessary CPD points. This exception is allowed only if 70 percent of the points were accumulated in the first 24 months of the certification period.

B. The recertification decision will be issued by NICET before the certification expires. The only exceptions are:

1) Additional time required for NICET to complete an audit of information supplied.

2) Additional time required by NICET for an upgrade denial to be reviewed or an upgrade decision to be made.

10. Candidates who successfully complete the recertification requirements will be issued a certificate and wallet card showing the new certification expiration date.

11. Failure to accumulate the required CPD points before the end of the certification period will cause the certification to expire and the certificant's listing of that certification in the registry will be deleted.

A. Only one notice of the need to reinstate an expired certification will be mailed to the certificant at the end of the first month that follows the end of the certification period.

12. Reinstatement of an expired certification to Active Status can only occur during the two years immediately after the expiration date. The return to Active Status requires the certificant to:

A. Acquire the 90 CPD points required for the certification period, which ended with the certification being placed in Expired Status.
B. Acquire an additional 30 CPD points if reinstatement is sought in the first twelve months of Expired Status or an additional 60 CPD points if reinstatement is sought in the second twelve months of Expired Status.

C. Pay the recertification fee and a reinstatement fee.

Note: Successful reinstatement of an expired certification to Active Status places the certification into the 3-year time period that would have gone into effect if the certification had not expired. For example, if the original certification expired on 7/1/2000 — the reinstated certification will expire on 7/1/2003.

13. An expired certification that is no **longer eligible for reinstatement** can be returned to Active Status only by reapplying as an initial (first time) applicant and satisfying the current certification criteria.

14. Inactive Status may be requested at anytime during the certification period and can be held for a maximum of 36 consecutive months. Inactive Status interrupts the three-year certification period and thus moves the expiration date of the certification(s) to a later date. Conditions for Inactive Status are:

A. Certificant becomes involuntarily inactive in certification practice area due to circumstances beyond his/her control, such as extended medical leave, extended military duty or extended jury duty.
B. All certificates held must go into Inactive Status.

C. Inactive Status cannot be requested until after the period of inactivity in the certification practice area(s) equates to what would have been 500 hours of normal employment for the certificant's peers.

D. Inactive Status must be requested in writing within 12 months of the start of the period of inactivity in the certification area(s).

E. The certification(s) will automatically expire unless returned to Active Status.

1) Return to Active Status must be accomplished within 12 months of the end of the period of inactivity and requires payment of a reinstatement fee.

2) Inactive Status certification(s) which expire can be regained only by reapplying as an initial (first-time) applicant and satisfying the current certification criteria (see Item #13 above).

15. Retired Status may be elected by certificants with Active Status certification(s) at the end of the certification period if the certificant retires from all paid employment in his/her certification practice area(s).

A. Retired Status requires the submission and acceptance of the recertification application, proof of retirement (written statement from employer), and payment of the Retired Status fee.

B. Retired Status is available only for the three-year period immediately following the Active Status certification period.

1) During this three-year Retired Status period, the Retired Status certificant will receive the NICET Newsletter and will appear in all appropriate NICET directories with certification(s) noted as "Retired."

2) At the end of the three-year Retired Status period, all certifications of record will expire and all test history will be deleted from the database. Regaining the expired certificate(s) is only possible by reapplying as an initial (first-time) applicant and satisfying the current certification criteria (see Item #13 above).

3) After the three-year Retired Status period, the NICET Newsletter will be available through the payment of a newsletter subscription fee.

C. Retired Status will be assigned to all certificates held.

D. Return to Active Status can be accomplished upon meeting the following conditions:

1) Obtained in accordance with Section II Exhibit I. Certifications not returned to Active Status will automatically expire.

2) Payment of the recertification fee and a reinstatement fee.

16. Implementation of this policy began on January 1, 1996.

A. All active certificate holders were assigned a three-year certification expiration date.
B. All first-time certifications issued July 1, 1997 or later will carry an expiration date of three years after the issue date.

Note 1: A certification practice area encompasses those activities <u>associated with the certification awarded</u>. For example, if you are certified in Highway Construction -- <u>that</u> is your certification practice area. If you are certified in Fire Alarms Systems -- <u>that</u> is your certification practice area. If you are certified in Industrial Engineering Technology -- <u>that</u> is your certification practice area.

Exhibit I.

Continuing Professional Development Points
(Revised October 1997)

I. Point Assignment to Continuing Professional Development Activity Categories:

A. Total points required in the 3-year period for each certification (see Notes 1 & 2) 90 (100%)

B. Maximum points possible in each activity category

    1. Active Practitioner/up to 24 points per year (72 points in 3 years) (80%)
    2. Additional Education/up to 45 points in 3 years (50%)
    3. Advance Profession/up to 45 points in 3 years (50%)
    4. Certification Activity/up to 45 points in 3 years (50%)
    5. Pass Special Exam/60 points (67%)

Note 1: When seeking recertification in more than one certification practice area, 90 points are needed for each certification. Dual counting of points is permissible in most categories where job assignments or point accumulation activities can reasonably be considered as useful or common to multiple certification practice areas (refer to Item #5 of Policy 30). For example, one may be an active practitioner in more than one certification practice area during an employment period and may earn up to 24 points per year <u>for each certification.</u>

Note 2: The intent of Policy #16, "Foreign Work Experience," applies to all facets of the recertification policy. The foundation of the NICET certification process is based upon work experience gained in the United States. Equally, the preponderance of CPD points for recertification must be acquired in the United States. This rule applies to "active practitioner" work experience and to all other point-accumulating activities (Additional Education, Advance Profession, and Certification Activity).

II. Point Determination Within Specific Activity Category:

A. ACTIVE PRACTITIONER

An active practitioner is defined as a certificant whose primary employment results in significant time periods performing tasks that are directly associated with the technical aspects of the certification practice area(s). The certificant may function as a hands-on practitioner, as an immediate supervisor, or as a technical project manager. Primary employment refers to the certificant's main job and is not dependent upon remaining with the same employer.

Holding a regular second job (moonlighting) does not provide any active practitioner points even if the second job involves a different certification practice area.

    1. Point Earning (Maximum of 24 in 1-year period):

        a. 24 points for the certification practice area when primary employment equals or exceeds 1000 hours or 120 workdays (whichever occurs first) and involve relevant tasks/activities that occur on a regular basis during the employment period. (A workday is defined as 7 hours or more.)

        b. When primary employment is less than 1000 hours or 120 workdays, the points earned are calculated by one of the following formulas:

$$\frac{\text{number of hours worked} \times 24}{1000 \text{ hours}}$$

or

$$\frac{\text{number of days worked} \times 24}{120 \text{ days}}$$

Points claimed by these formulas are to be whole numbers only (round up allowed for all decimals). **No CPD credit is earned when the calculated results are 7 points or less**.

B. ADDITIONAL EDUCATION

Additional education is defined as efforts undertaken to advance, broaden, and enhance the certificant's technical knowledge and job skills. These efforts may range from taking a college course for a grade to a less formal company sponsored in-house training session. NICET will not pre-approve courses, seminars or other training efforts for CPD points. It is the responsibility of the certificant to assure that such activities meet the NICET requirements, particularly the requirement that the additional education constitutes **professional development** for that individual; in other words, it must **increase the certificant's knowledge** — not merely cover what should already be known.

    1. Point Earning (Maximum of 45 in the 3-year period):

        a. College credit courses:

            1) 15 points per semester hour;
            2) 10 points per quarter hour.

b. Offerings with preassigned CEUs:

> **1)** 1 point per 0.1 CEU (10 points per CEU).

c. Less formal offerings such as workshops, seminars, technical presentations at meetings, and training sessions:

> **1)** 1 point per each contact hour.
> A contact hour is defined as the 45 to 60-minute period devoted to the learning effort. Fractional points are not recognized — a stand-alone 30-minute session has no point value, however a course of two 30-minute sessions (even if sessions are on different days) has a value of one point.

C. ADVANCE PROFESSION

Advance profession is defined as volunteer or for-hire activities such as providing certification practice area expertise to assist committees, task forces, and the like to reach their objectives; making presentations at meetings of others who seek to increase their knowledge of the certification practice area; writing technical research or position papers pertinent to the certification practice area for publication/distribution; serving as an instructor for a training/educational offering pertinent to the certification practice area; participating actively in career day, science fair and related events; writing acceptable NICET certification examination questions; serving on an educational institution curriculum development advisory committee; and participating in professional membership organizations which contribute to the advancement of the certification practice area or the professional role of the engineering technician and the engineering technologist.

1. Point Earning (Maximum of 45 in the 3-year period):

a. Active Committee/Task Force Service

An active committee/task force is defined as one conducting technical business by way of activities such as one or more meetings per year, one or more reports/position papers issued per year, one or more studies/surveys underway per year, etc. Service on an educational institution curriculum development advisory committee is considered to be relevant committee/task force service, as is serving as a NICET certification examination question writer/reviewer. Point earning is as follows:

> **1)** 1 point per month as chair or co-chair of an active national committee/task force; 0.75 point per month for an active state/regional committee/task force, and 0.5 point per month for an active local committee/task force.
> **2)** 0.5 point per month as a participating member of an active national committee/task force, 0.375 point per month for an active state/regional committee/task force, and 0.25 point per month for an active local committee/task force.

3) 0.6 point per each NICET certification exam question written and accepted; 0.4 point per each NICET certification exam question reviewed and accepted; up to a maximum of 15 points during the three-year period.

b. Presentations

A presentation is defined as the preparation and subsequent audio, visual, and/or written delivery of technical information or information about the profession to others who are seeking to advance, broaden, and enhance their knowledge and/or job skills. Point earning is as follows:

1) 20 points for being the sole author (15 points for being a co-author) of a published refereed technical paper or instructional materials for a course.

2) 10 points for being the sole author (5 points as a co-author) of a published non-refereed technical paper or instructional materials for a course.

3) 15 points for being the sole author (8 points as a co-author) of a published refereed profession explanation/position paper.

4) 8 points for being the sole author (4 points as a co-author) of a published non-refereed profession explanation/position paper.

5) 10 points for delivery of your published paper to an audience at a scheduled meeting.

6) 5 points for delivery of your non-published paper to an audience at a scheduled meeting.

c. Course Instructor

A course instructor is defined as a person who meets with and directs a technical educational effort designed for a particular type of student. (A first time offering of a course by an instructor is also covered under 1) and 2) of the "Presentations" category above.) Point earning is as follows:

1) 0.7 point per each contact hour when conducting a training course which has not been preassigned CEUs.

2) 1 point per each contact hour when conducting a training course with preassigned CEUS.

3) 15 points per semester hour (10 points per quarter hour) when serving as the instructor for a scheduled course in a post-secondary educational institution.

d. Career Day Presenter A career day presenter is defined as a person who meets with and explains to students the various aspects of careers in engineering and engineering technology; serves as a judge at a science/engineering fair; is an active volunteer in an Engineers Week event; is an active volunteer in a MATHCOUNTS contest; or actively participates in other related activities. Point earning is as follows:

1) 1 point per participation day, to a maximum of 10 points during the three-year period.

e. Professional Society Activity

Professional society activity is defined as involvement with organizations that exist to promote technical knowledge and techniques pertinent to the certification practice area or that exist to unify individuals in a particular career area. Involvement is further defined as activities, which are not specifically covered by the "Active Committee/Task Force Service" category above. Point earning is as follows:

1) 1 point for each year of a society membership held (up to 3 points per society), to a maximum of 12 points. (Caution - a local and a national membership **in the same society** can only provide 3 points.)

2) 1 point for each local or state meeting attended, to a maximum of 10 points.

3) 2 points for each regional or national meeting attended, to a maximum of 12 points.

4) 2 points for each year of service as an elected regional or national officer, to a maximum to 12 points.

5) 1 point for each year of service as an elected local or state officer, to a maximum of 12 points.

6) 1 point for each year of service in an appointed position at the regional or national level, to a maximum of 6 points.

7) 0.5 point for each year of service in an appointed position at the local or state level, to a maximum of 6 points.

D. CERTIFICATION ACTIVITY

Certification activity is defined as the efforts required to upgrade a NICET certification and/or achieve a certification (NICET or non-NICET) in a related practice area

1. Point earning is as follows (maximum of 45 in the 3-year period):

a. upgrade in the certification practice area:

1) 35 points for achieving an upgrade of the NICET certification during the 3-year certification period

2) 10 points for meeting a higher level exam requirement (but certification not achieved) during the 3- year certification period; or

3) 1 point for each NICET work element passed during the 3-year certification period that can be used to meet the higher level exam requirement (but full exam requirement not met).

b. certification in a related practice area:

1) 25 points for each NICET certification achieved (initial or upgrade) during the 3-year certification period:

a) 8 points for meeting a higher level exam requirement (but certification not achieved) during the 3- year certification period; or

b) 0.5 points for each NICET work element passed during the 3-year certification period that can be used to meet the higher level exam requirement (but full exam requirement not met).

2) 15 points for each NICET-recognized, non-NICET certification achieved during the 3-year certification period.

c. excess work element in the certification practice area: these work elements must be in excess of those required to meet all higher level exam requirements (through Level IV) and cannot, at any time, be counted under Item D.1.a.3) above.

1) 1 point for each existing NICET work element passed during the 3-year certification period
2) 1 point for each newly-available NICET work element passed during the 3-year certification period through participation in a field test or through regular testing within 2 years of the introduction of the new work elements

E. SPECIAL EXAM

The special recertification exam is defined as a 3-hour written examination designed specifically for the actual certification practice area and certification level (grade).

1. Point earning is as follows (maximum of 60 in the 3-year period):

a. 60 points for passing the appropriate NICET special examination for the certification practice area.

POLICY 31. SUSPENDED CERTIFICATION STATUS
(NEW POLICY - EFFECTIVE MARCH 1998)

Suspended Status can result from Policy #1 or Policy #2 actions. During the suspension time period, all existing certifications, pending certifications, and testing activity will be frozen. The accumulation of CPD points for the scheduled recertification (Policy #30) is not waived during the suspension period.

Step 2: Review the Four Certification Criteria

1 - Relevant work experience.

Sufficient work experience relevant to the desired certification is necessary to qualify for a particular level of certification. The amount of experience required for each level is:

Level I - several months

Level II - 2 years minimum

Level III - 5 years minimum

Level IV - 10 years minimum

Your work experience must be progressively more technical and responsible.

2 - Written examination

You may select work elements for testing according to your personal employment situation and experience. Although it is possible to meet the Level II certification exam requirement in just one sitting, about half of all applicants need a second examination to finish the requirement.

3 - Supervisor Evaluation of On-The-Job Performance

Your immediate supervisor is required to affirm that he/she has personally observed you repeatedly and correctly perform the task(s) and/or apply the knowledge associated with the particular work element under a variety of on-the-job conditions.

4 - Recommendation from a Qualified Individual

A personal recommendation is required from a professional who is familiar with your technical capabilities and background, including the quantity and quality of your work experience.

Step 3: Read the Program Detail Manual

The Fire Alarm Systems certification program became operational in 1988. It was designed for engineering technicians working in the fire alarm industry who are engaged in a combination of the following fire alarm systems activities:

System Layout (plans preparation)	System trouble-shooting
System equipment selection	System servicing
System installation	System sales
System acceptance testing	

Technical areas covered include applicable codes and standards; types of signaling systems; supervision requirements; types of fire and smoke detectors; building occupancy considerations; basic electricity and electronics; and physical science fundamentals.

Field code ID numbers

A 3-digit number identifies each technical field. The technical field code for Fire Protection Engineering Technology is 003.

The identification number assigned to each work element is 5 digits long:

The first digit identifies the technical sub-field within the field of Fire Protection Engineering Technology as follows:

1 - Automatic Sprinkler Systems Layout

2 - ~~Special Hazards Systems Layout~~ (no longer available)

3 - Fire Alarm Systems

4 - Inspection and Testing of Water-Based Fire Protection Systems

5 - Special Hazards Suppression Systems

The second digit identifies the level and the work element type as follows:

1-Level 1 General	**3**-Level II General	**5**-Level III General	**7**-Level IV General
2-Level I Special	**4**-Level II Special	**6**-Level III Special	**8**-Level IV Special

Each element in the Program Detail Manual includes a description of the material covered by the exam and any references used such as NFPA publications.

Step 4: Review the examination requirement chart

You must pass these exam elements to complete the **Level I** exam requirement.

Level I – General	6
Level I – Special	2
Total	**8**

You must pass these work elements to complete the **Level II** exam requirement. **Read note (a).**

Level I – General	6a
Level I – Special	2
Level II – General	18
Level II – Special	4
Total	**30**

You must pass these work elements to complete the **Level III** exam requirement. **Read notes (a), (b), and (c).**

Level I – General	9c
Level I – Special	3
Level II – General	24c
Level II – Special	6
Level III – General	11c
Level III – Special	1
Total	**54**

You must pass these work elements to complete the **Level IV** exam requirement. **Read notes (a), (b), (c), and (d).**

Level I – General	9c
Level I – Special	3
Level II – General	26c
Level II – Special	6
Level III – General	15c
Level III – Special	1
Level IV – General	6c
Level IV – Special	2
Total	**68**

NOTES:

(a) Work Element #31011, "Basic Metric Units and Conversion," must be passed to achieve certification at Levels II, III, and IV.

(b) Time restrictions dictate that no more than 34 work elements can be scheduled for any single examination sitting. Therefore, at least two examination sittings will be needed in order to complete this requirement.

(c) All core work elements in this category must be passed to complete the exam requirement at this level.

(d) Read very carefully the two sections applicable to Level IV certification in the PDM before seeking Level IV certification.

GENERAL NOTES:

Work elements passed which are in excess of the exam requirement for a particular type and level, but which are needed to meet the requirement at the next higher level are automatically applied to that higher level requirement.

Use the Personal Tally Worksheet on page 19 of the PDM to keep track of the number of work elements you have successfully passed.

Step 5: Technician Personal Recommendation Form

Complete the header information and ask your recommender to complete the form.

The recommender:

- Must be qualified (PE, NICET, senior technician, etc.).

- Must be familiar with applicant's capabilities.

- May not be a relative or of a non-technical background.

There is no deadline for this form. The Technician Personal Recommendation Form will not be considered until the exam requirement is met.

Step 6: Complete the Technician Application Form

Part I   **Personal Information**
Your name, address, phone number, etc.

Part II   **Employment Information**
Your employer's name, address, phone number, etc.

Part III   **Applicant Information**
Whether or not you are testing and / or are certified.

Part IV   **Examination Scheduling**
Indicate your 1st and 2nd choice of exam location and date.

Part V   **Education Information**
Focus on relevant education information.

Part VI   **Employment History**
Follow the instructions and describe your relevant job duties.

Part VII   **Work Element Verification**
Your immediate supervisor must affirm your job knowledge.

Part VIII   **Work Element Examination**
Carefully select then list each element by number and name.

Part IX   **Part A / Part B Examination**
Leave this section blank.

Part X   **Applicant's Statement of Understanding**
Read the statement of understanding and affirm your agreement by signing
your name.

Step 7: Determine your test center location and date

Test Center Code __________ **at**____

	Cycle 1	Cycle 2	Cycle 3	Cycle 4
Exam Date				
Mail Deadline				

Test Center Code __________ **at**____

	Cycle 1	Cycle 2	Cycle 3	Cycle 4
Exam Date				
Mail Deadline				

Test Center Code __________ **at**____

	Cycle 1	Cycle 2	Cycle 3	Cycle 4
Exam Date				
Mail Deadline				

Test Center Code __________ **at**____

	Cycle 1	Cycle 2	Cycle 3	Cycle 4
Exam Date				
Mail Deadline				

Test Center Code __________ **at**____

	Cycle 1	Cycle 2	Cycle 3	Cycle 4
Exam Date				
Mail Deadline				

Step 8: Submit your application

Technician examination fee   **$150.00**
When applying to test, include the examination fee and mail your application as soon as possible to improve your chance of reserving a seat for the exam.

Annual registry fee   **$30.00**
Certification is granted for a 3-year period. The registry fee is payable annually thereafter. Your notice to renew for the 3rd year will include a Continuing Professional Development form. You are required to account for 90 CPD points (see policy #30) before your certificate will be renewed for another 3-year term.

Send your exam application and fee to:

**NICET
c/o NationsBank
Dept 0037
Washington, DC 20055**

Step 9: Practice Tai Chi to relax and improve patience

NICET will grade your exam and you will receive the results in about 10-14 days. Once you have successfully passed the proper number and types of exam elements, NICET will begin to corroborate the information you provided on your application and personal recommendation form. This will take 90 to 120 days. If additional information is required, you will receive a conditional certification letter requesting additional work experience information, personal recommender(s), and/or other information necessary before certification is bestowed upon you.

All good things take time so relax and practice Tai Chi. This will also help you wind down after a rough day with the inspector.

What to expect on exam day

Arrive and check in before 8am. At approximately 8:30am the proctor will read the rules and announcements. You will begin your exam at approximately 9am. If your test is longer than 3 hours, you are required to leave the classroom for a one-hour lunch break.

Things to bring with you:

- The confirmation notice you receive from NICET along with picture ID.

- At least two #2 pencils w/ eraser plus highlighters of two different colors.

- Calculator with batteries (no laptop computers or PDA's).

- Code books, handbooks, standards, references, and manuals, which must be bound or contained in a 3-ring binder. Loose sheets of paper are forbidden.

The Exam

- Fan-fold test questions and a separate Scantron® answer sheet.

- Expect 5-10 questions per element and 1-2 minutes to answer each question.

- All questions are multiple-choice. You must choose the most correct answer.

- Be prepared to break for 1 hour at lunchtime and leave the testing room.

- You must correctly answer at least 60% of the questions to pass an element.

- Use your time wisely and <u>answer all questions</u>, even if you must guess.

Test-Taking Tactics

1. Begin reading the test questions. If you experience "brain-lock", continue to read test questions until you can confidently answer from memory.

2. Highlight the test questions you are unable to answer quickly either from memory or by quickly confirming your choice by referring to your codes, standards, and handbooks.

3. Continue reading test questions, highlighting those that you can't answer quickly (1-2 minutes) until you read all test questions. You should be able to accomplish this "first pass" before the lunch break.

4. After the lunch break, begin re-reading the test questions you highlighted before lunch, spending more time to look up answers in your codes, standards, and handbooks. Be careful not to get mired in a complicated question you are not at all familiar with.

5. You should be able to complete this "second pass" with at least 30 minutes remaining before the scheduled end of your test. Use the second highlighter color to mark those questions you haven't a clue about. The time at which you will be expected to finish your exam should appear on the blackboard or whiteboard at the front of the classroom.

6. After your "second pass", begin to re-read the questions you highlighted with the second color and guess. Often times you can eliminate two answers as obviously incorrect leaving a 50-50 chance at guessing the correct answer.

7. With the remaining few moments before the end of your exam, review your answer sheet to ensure that the header information and all questions have been answered.

I've applied and taken my exam, now what?

A score report will be mailed to you within 18 days at which time your work experience, personal recommendation, and supervisor evaluation will be considered.

Your certification will be determined within 90 days. If you do not receive a notice within 120 days, contact NICET at (888) IS-NICET.

Once certified you will receive an award letter, certificate, wallet card, and six-digit ID number, at which point you may identify yourself with the suffix CET or;

Level I TT (Technician Trainee)   Level III ET (Engineering Technician)
Level II AET (Associate Engineering Technician)   Level IV SET (Senior Engineering Technician)

Once you're certified, follow policy #30 (Continuing Professional Development), prepare for the next exam cycle, and practice the secret handshake.

'Greasing the Skids'

NICET knows that waiting 90 days after your exam to see whether you have been approved for certification is a very long time to wait. The 90-day wait can have a happier outcome if you send them everything they need to evaluate your eligibility before you test.

> <u>When your file is reviewed, your evaluator will be looking for 3 things:</u>
>
> 1. A good work history.
>
> 2. A current personal recommendation from an appropriate person
>
> 3. Your supervisor's verification (initials) of work elements

- Be sure your work history is detailed, informative, and complete. Don't write two sentences to describe three years of work. Give NICET the names of all supervisors, past and present, your job title changes, and the dates of employment at your current and your previous employers.

- Personal Recommendations stay on file for one year from the signature date. Remember that the same person cannot recommend you and verify your work elements.

- Remember also that the person whom you have identified as your current supervisor should be the person you use to verify your elements. If you don't have a supervisor (or if your supervisor is not knowledgeable in your technical area), use another person who has been in a position to supervise, inspect or approve your work and explain to NICET at what point in time that person was in such a position.

- Candidates for Level IV have an additional requirement to submit a major project write-up. Guidelines for this write-up can be found in the current edition of the Program Detail Manual.

If you have left out any of these things or if the work history provided is not sufficiently detailed or complete, your certification will be delayed and NICET will send you a "Conditional Decision Letter" asking for additional information. Once you return the required information, your file will wait in line for a second review.

Additional Resources

1. The NICET wesite http://www.nicet.org includes:
   - Application forms and addenda
   - Program Detail Manuals
   - Exam dates

2. Call NICET at 888-IS-NICET.

3. Use your newsreader (Outlook Express, Agent, etc.) to participate in Usenet discussion at:
   - alt.engineering.fire-protection
   - alt.engineering.fire-alarm

   If you don't see these newsgroups on your list of newsgroups, email your ISP at news@yourisp.com [substitute the name of your internet service provider for 'yourisp'] and ask that they add these groups to their Usenet server.

4. Join the AFAA (Automatic Fire Alarm Association) list-server by visiting their website http://www.afaa.org and following the sinstructions.

5. Find out when NICET test preparation and other seminars are scheduled in your area by calling Limited Energy Resource Center at 503-657-0135.

6. Join the preeminent standard-writing authority NFPA, at http://www.nfpa.org. The cost is $115 per year, which gives you a voice in the standards-writing cycle and a 10% discount on NFPA publications. It's tax deductible!

7. Point your browser to my website at http://go.to/baker where you will find lots of cool fire alarm crud including:
   - This guide
   - Volt-drop worksheet & Excel spreadsheet
   - Power-Point presentations
   - Much more!

8. Contact me, if you prefer, using one or more of the following:
   - Voice _______________________________
   - Cell _______________________________
   - Fax _______________________________
   - email Michael.B.Baker@home.com

Notes

SMALLER WIRE SIZES - Conductor Properties

Size	20	20	22	22	24	24
# Strands	1	7	1	7	1	7
Cir Mils	1020	1113	643	700	404	448
Res kft	10.695	10.485	16.995	16.895	27.682	26.167

Size	26	26	28	28	30	30
# Strands	1	7	1	7	1	7
Cir Mils	253	278	159	175	100	112
Res kft	45.29	43.255	72.058	68.699	108.494	97.909

VOLTAGE DROP NOTES & FORMULA

I = amps   **R** = resistance   **VD** = voltage drop

VD permitted = voltage drop allowed by NEC on any branch circuit (3%) (120v x 3% = 3.6v)

CM = size of conductor in circular mils *(Table 8)*

D = distance of the circuit one way

K = is resistance of a circular mil-foot of wire *(Table 8)*

Ⓚ = approximate K   copper = 12.9, aluminum = 21.2

PL = power loss (wasted electricity) due to resistance of conductors

FORMULAS

Exact K = $\dfrac{R \times CM}{1000'}$ **Power Loss** = VD x I

Voltage Drop $VD = \dfrac{2 \times K \times D \times I}{CM}$ or   VD = I x R

Wire Size $CM = \dfrac{2 \times Ⓚ \times D \times I}{VD \text{ permitted}}$

Distance $D = \dfrac{CM \times VD \text{ permitted}}{2 \times K \times I}$

Load $I = \dfrac{CM \times VD \text{ permitted}}{2 \times K \times D}$

LIMITED ENERGY RESOURCE CENTER

CEU, LICENSING PREP COURSES & CUSTOMIZED EDUCATION COURSES

Phone: (503) 657-0135   Fax: (503) 657-2911

Level I Element Review

Element	Title	Reference Documents
31001	Basic Fire Alarm Systems	NFPA 72 National Fire Alarm Code NFPA 101 Life Safety Code Fire Alarm Signaling Systems
31002	NFPA Standards	Definition and Scope Sections of Standards
31003	Basic Wiring	NFPA 70 National Electrical Code NFPA 72 National Fire Alarm Code UL Electrical Construction Materials Directory
31004	Devices & Components	NFPA 72 National Fire Alarm Code Training Manual on Fire Alarm Systems Fire Protection Handbook Fire Alarm Signaling Systems
31005	Periodic Tests	NFPA 72 National Fire Alarm Code
31006	Basic Electricity	
31007	Basic Working Drawings	NFPA 72 National Fire Alarm Code NFPA 170 Fire Safety Symbols Drafting texts
31008	Basic Mathematics	General mathematics textbooks
31009	Installation Practices	NFPA 72 National Fire Alarm Code
31010	Basic Communication Skills	Basic grammar references
31011	Metric Units & Conversions	ASTM E-380
32001	Plans, Specifications, & Contracts	General construction texts and references
32003	Basic Physical Science	Solutions may involve simple formulas found in basic physics textbooks, but will not involve algebraic manipulations or trigonometry.
32004	Fire Warning Equipment for Dwelling Units	NFPA 72 National Fire Alarm Code
32005	Basic Individual Safety	OSHA 2201 General Industry Digest
32006	First Aid Procedures	General handbooks on first aid

Notes

31001 BASIC FIRE ALARM SYSTEMS

REFERENCES:

NFPA 72

NFPA 101

*Fire Alarm
Signaling Systems*

DESCRIPTION:

Understand the various types of fire alarm systems. Understand the electrical requirements, the alarm initiating devices, the control functions, the alarm indicating appliances, and the power requirements of a fire alarm system. Know the types of signaling services that can be provided and the automatic fire detectors in common use.

Why are fire alarm systems installed?

- Life safety

- Property protection

- Mission continuity

- Heritage preservation

- Environmental protection

NFPA 72-1999 1-3.1

Fire alarm system classifications

1) Household fire warning systems

2) Protected premises fire alarm systems

3) Supervising station fire alarm systems

    a) Auxiliary fire alarm systems

    b) Remote supervising station fire alarm systems

    c) Proprietary supervising station systems

    d) Central station fire alarm systems

    e) Municipal fire alarm systems

Components of a fire alarm system include:

- Fire Alarm Control Panel
    - To provide power
    - To monitor circuit integrity
- Initiating Devices
    - Manual (pull station, guard's tour)
    - Automatic (smoke, heat, waterflow, flame detector)
- Notification appliances
    - Audible (bell, horn, chime)
    - Visible (strobe)
    - Tactile / Olfactory (bed shaker / smoke)
- Transmitters and receivers
    - DACT ➠ DACR ('dialer')
    - McCulloh
- Control functions
    - Elevator recall
    - Fan shutdown

NFPA 72-1999 1-1

The *National Fire Alarm Code* covers the application, installation, location, performance, and maintenance of fire alarm systems and their components.

Alarm initiating is either from manual fire alarm boxes or automatic fire detectors and sensors.

The fire alarm control unit is the system controller. It provides power to the system and electrically monitors all circuits extending from it. The control unit switches power to notification appliances (horns, bells, strobes, etc.), building fire safety controls (fan shutdown, elevator recall, elevator shunt, etc.), and supplementary equipment based upon input from initiating devices (smoke detectors, heat detectors, manual pull boxes, etc.).

NFPA 72-1999 1-5.2.3

Fire alarm systems shall be provided with two independent and reliable power supplies of adequate capacity for the application.

Initiating device types include conventional, conventional-addressable, and analog-addressable.

Individual conventional detectors include the technology within them to determine whether an alarm condition exists then transmit the alarm to the fire alarm control panel by causing an increase in circuit current. The condition is annunciated at the control panel by circuit (or zone), not by device.

NFPA 72-1999 1-5.4.2.1 A coded alarm signal shall consist of not less than three rounds of the number transmitted and each round shall consist of no less than three impulses.

Conventional-addressable devices add the ability to annunciate by device.

Addressable-analog devices reduce individual detectors to sensors, allowing the fire alarm control panel to decide whether an alarm condition exists, then annunciate the condition by device.

NFPA 72-1999 1-5.6 Automatic smoke detection shall be provided at the location of each fire alarm control unit unless the location is continuously occupied.

NFPA 72-1999 1-5.8.1 All circuits and pathways interconnecting fire alarm system components shall be monitored for integrity.

NFPA 72-1999 1-6.2.1 A record of completion shall be prepared for every fire alarm system.

NFPA 72-1999 1-6.3 Records of fire alarm system tests and operations shall be kept until the next test and for 1 year thereafter.

NFPA 72-1999 3-4.2 Class or style, or both may be used to indicate the level of fault tolerance in initiating, notification, and signaling line circuits.

NFPA 72-1999 3-4.2.1(1) Circuits capable of transmitting an alarm signal during a single open or a non-simultaneous single ground fault shall be designated as Class A.

NFPA 72-1999 3-4.2.1(2)

Circuits not capable of transmitting an alarm beyond the location of a single open or non-simultaneous single ground fault shall be designated as Class B.

NFPA 72-1999 3-4.2.2.1(1)

NFPA 72-1999 Table 3-5

An initiating device circuit shall be permitted to be described as either Style A, B, C, D, or E, depending upon its fault tolerance, during a single open, single ground, wire-to-wire short, and loss of carrier fault condition.

NFPA 72-1999 3-4.2.2.1(2)

NFPA 72-1999 Table 3-6

A notification appliance circuit shall be permitted to be designated as either Style W, X, Y, or Z, depending upon its fault tolerance, during a single open, single ground, and wire-to-wire short condition.

NFPA 72-1999 3-4.2.2.1(3)

NFPA 72-1999 Table 3-7

A signaling line circuit shall be permitted to be designated as either Style 0.5, 1, 2, 3, 3.5, 4, 4.5, 5, 6, or 7 depending upon its fault tolerance, during a single open, single ground, and wire-to-wire short condition.

NFPA 101 1997 1-2.3

NFPA 101 *Life Safety Code* identifies criteria required for the design of egress facilities permitting prompt escape of occupants from buildings.

NFPA 101-1997 7-6.1.8

NFPA 101-1997 7-7.6

Where a required fire alarm system or automatic sprinkler system is out of service for more than 4 hours in a 24-hour period, the AHJ shall be notified and the building evacuated or an approved fire watch shall be provided.

NFPA 101-1997 7-6.7.3

For the purposes of annunciation, each floor of a building shall be considered a zone.

NFPA 101-1997 7-6.7.4

Where a floor area exceeds 20,000 ft^2, additional zoning shall be provided. The length of any zone shall not exceed 300 ft.

Review the glossary of this Handbook for definitions.

31002 NFPA STANDARDS

DESCRIPTION:

Understand the basic application of NFPA standards to fire alarm systems. Understand the basic NFPA terminology including "will" and "should" and the role of the "authority having jurisdiction". Understand the basic concept of "approved", "listed", etc. in regard to the acceptance of materials, components, and devices and the role of testing laboratories in relation to the use of NFPA standards. Select and use appropriate NFPA standards.

Glossary of Terms

Approved. Acceptable to the authority having jurisdiction. The National Fire Protection Association does not approve, inspect, or certify any installations, procedures, equipment, or materials; nor does it approve or evaluate testing laboratories. In determining the acceptability of installations, procedures, equipment, or materials, the authority having jurisdiction may base acceptance on compliance with NFPA or other appropriate standards. In the absence of such standards, said authority may require evidence of proper installation, procedure, or use. The authority having jurisdiction of an organization concerned with product evaluations that are in a position to determine compliance with appropriate standards for the current production of listed items.

Authority Having Jurisdiction. The organization, office, or individual responsible for approving equipment, an installation, or a procedure. The phrase "authority having jurisdiction" is used in NFPA documents in a broad manner, since jurisdictions and approval agencies vary, as do their responsibilities. Where public safety is primary, the authority having jurisdiction may be a federal, state, local, or other regional department, or individual such as a fire chief; fire marshal, chief of a fire prevention bureau, labor department, or health department; building official; electrical inspector; or others having statutory authority. For insurance purposes, an insurance inspection department, rating bureau, or other insurance company representative may be the authority having jurisdiction, In many circumstances, the property owner or his or her designated agent assumes the role of the authority having jurisdiction; at government installations, the commanding officer or departmental official may be the authority having jurisdiction.

Fire Alarm Signal. A signal initiated by a fire alarm-initiating device such as a manual fire alarm box, automatic fire detector, waterflow switch, or other device whose activation is indicative of the presence of a fire or fire signature.

Fire Alarm System. A system or portion of a combination system consisting of components and circuits arranged to monitor and annunciate the status of fire alarm or supervisory signal-initiating devices and to initiate the appropriate response to those signals.

Labeled. Equipment or materials to which has been attached a label, symbol, or other identifying mark of an organization that is acceptable to the authority having jurisdiction and concerned with product evaluation that maintains periodic inspection of production of labeled equipment or materials and by whose labeling the manufacturer indicates compliance with appropriate standards or performance in a specified manner.

Listed. Equipment, materials, or services included in a list published by an organization acceptable to the authority having jurisdiction and concerned with evaluation of products or services that maintains periodic inspection of production of listed equipment or materials or periodic evaluation of services and whose listing states either that the equipment, material, or service meets identified standards or has been tested and found suitable for a specified purpose. The means for identifying listed equipment may vary for each organization concerned with product evaluation, some of which do not recognize equipment as listed unless it is also labeled. The authority having jurisdiction should utilize the system employed by the listing organization to identify a listed product.

Shall. Indicates a mandatory requirement.

Should. Indicates a recommendation or that which is advised but not required.

Spacing. A horizontally measured dimension related to the allowable coverage of fire detectors.

Subscriber. The recipient of contractual supervising station signal service(s). In case of multiple, noncontiguous properties having single ownership, the term refers to each protected premises or its local management.

Zone. A defined area within the protected premises. A zone can define an area from which a signal can be received, an area to which a signal can be sent, or an area in which a form of control can be executed.

31003 BASIC WIRING

REFERENCES:

NFPA 70

NFPA 72

UL Electrical Construction Materials Directory

DESCRIPTION:

Understand the wiring requirements and protection of wiring used for fire alarm systems. Select outlet and junction boxes, cable, and conduit. Calculate proper wire size and overcurrent protection for the system.

NFPA 70-1999 250-24(c) — A grounding electrode conductor shall be used to connect the equipment grounding conductors, the service-equipment enclosures, and, where the system is grounded, the grounded service conductor to the grounding electrode(s).

NFPA 70-1999 250-2(a) — Electrical systems required to be grounded shall be connected to earth in a manner that will limit the voltage imposed by lightning, line surges, or unintentional contact with higher voltage lines and that will stabilize the voltage to earth during normal operation.

NFPA 70-1999 250-2(d) — The fault current path shall be permanent and electrically continuous, shall be capable of safely carrying the maximum fault likely to be imposed on it, and shall have sufficiently low impedance to facilitate the operation of overcurrent devices under fault conditions.

Bonding shall be provided where necessary to ensure electrical continuity and the capacity to conduct safely any fault current likely to be imposed.

NFPA 70-1999 300-4 — Where subject to physical damage, conductors shall be protected.

NFPA 70-1999 300-11(a) — Raceways, cable assemblies, boxes, cabinets, and fittings shall be securely fastened in place. Support wires that do not provide secure support shall not be permitted as the sole support. Where independent support wires are used, they shall be secured at both ends. Cables and raceways shall not be supported by ceiling grids.

NFPA 70-1999 300-11(a)1

Where independent support wires are used, they shall be distinguished by color, tagging, or other effective means from those that are part of the fire-rated design.

NFPA 70-1999 300-13(a)

Conductors in raceways shall be continuous between outlets, boxes, devices, etc. There shall be no splice or tap within a raceway except as allowed by exceptions.

NFPA 70-1999 310-2(a)

Conductors shall be insulated.

NFPA 70-1999 310-2(b)

Conductors shall be of aluminum, copper-clad aluminum, or copper, unless otherwise specified.

NFPA 70-1999 760-27(a)

Only copper conductors shall be permitted to be used for fire alarm systems.

NFPA 70-1999 310-11(a)

All conductors and cables shall be marked to indicate the following information:

1) The maximum rated voltage.

2) The proper type letter or letters for the type of wire or cable.

3) The manufacturer's name, trademark, or other distinctive mark.

4) The AWG size or circular mil area.

5) Cable assemblies in which the neutral conductor is smaller than the ungrounded conductors shall be marked.

NFPA 70-1999

Table 310-13

Conductor Application and Insulation.

NFPA 70-1999 370-2

Round boxes shall not be used where conduits or connectors requiring the use of locknuts or bushings are to be connected to the side of the box.

NFPA 70-1999 370-4

All metal boxes shall be grounded.

NFPA 70-1999

Table 370-16(a)

See box-fill table for metal boxes in the NEC.

NFPA 70-1999 370-24

No box shall have an internal depth of less than 1/2 in.

NFPA 70-1999 370-25

In completed installations, each box shall have a cover, faceplate, or fixture canopy.

Review NFPA 70-1999 articles 250, 300, 310, 370, 402, 500, 725, and 760.

NFPA 70-1999 370-4

Notes

31004 DEVICES AND COMPONENTS

REFERENCES:

NFPA 72

Training Manual on
Fire Alarm Systems

Fire Protection Handbook

Fire Alarm
Signaling Systems

DESCRIPTION:

Understand the operation and use of manual fire alarm boxes, automatic fire detectors, audible and visible signaling appliances, annunciators, and other basic components of a fire alarm system.

Initiating devices for fire alarm systems are either manual fire alarm boxes or automatic detectors and are used to activate an alarm on a fire alarm system.

NFPA 72-1999 2-8.1

The operable part of each manual pull box shall not be less than 3 1/2 ft nor more than 4 1/2 ft above floor level.

NFPA 72-1999 2-8.2.2

Manual fire alarm boxes shall be located within 5 feet of the exit doorway opening at each exit on each floor.

NFPA 72-1999 2-8.2.3

Manual fire alarm boxes shall be mounted on both sides of group openings over 40 feet in width.

NFPA 72-1999 2-8.2.4

Manual fire alarm boxes shall be distributed throughout the protected area such that travel distance to the nearest fire alarm box will not be in excess of 200 ft measured horizontally on the same floor.

Manual fire alarm boxes may be:

1) Coded or non-coded

2) Presignal or general alarm

3) Breakglass or non-breakglass

4) Single or double action

Once actuated, a non-coded manual fire alarm box maintains the alarm condition until reset.

NFPA 72-1999 1-5.4.2.1   A coded alarm shall consist of not less than three complete rounds of the number transmitted, and each round shall consist of not less than three impulses.

Presignal fire alarm boxes initially cause alarm signals to sound only in specific areas. Actuation of a key switch on the fire alarm box or the control unit will cause an evacuation signal to sound.

A general alarm fire alarm box (the most common), when actuated, causes evacuation signals to sound immediately.

The term "breakglass" is applied to both non-coded and coded fire alarm boxes where the actuation of the device to cause an alarm requires the initial action of breaking a glass or other breakable element. Fire alarm boxes without this feature are classified as "non-breakglass".

A single action fire alarm box is a fire alarm box that initiates an alarm as a result of a single action by the user. The required action usually is breaking a glass element only or actuating a lever or other movable part of the station.

Automatic alarm initiating devices may be actuated by various factors that may be present as the result of a fire. These factors may be direct effects such as heat, smoke, flame radiation, or combinations of these effects.

Heat detector types include:

    1) Fixed-temperature

    2) Rate-compensated

    3) Line-type

A fixed temperature heat-sensing fire detector is a device that will respond when its operating element becomes heated to a predetermined level.

Typical examples of fixed temperature sensing elements are:

1) Bimetallic

2) Electrical conductivity

3) Fusible alloy

4) Heat sensitive cable

5) Liquid expansion

A rate compensation detector is a device that will respond when the temperature of the air surrounding the device reaches a predetermined level regardless of the rate of temperature rise.

A rate-of-rise heat detector is a device that will respond when the temperature rises at a rate exceeding a predetermined amount (usually 15 degrees Fahrenheit per minute).

Examples of rate-of-rise heat-sensing fire detectors include:

1) Pneumatic rate-of-rise tubing

2) Spot-type pneumatic rate-of-rise detector

3) Thermoelectric effect detector

4) Electrical conductivity-type rate-of-rise detector

Smoke-sensing fire detectors are classified by operating principle:

1) Ionization smoke detector

2) Photoelectric light-obscuration smoke detector

3) Photoelectric light-scattering smoke detector

4) Cloud chamber smoke detector

An ionization smoke detector uses a small amount of radioactive material to ionize the air in the sensing chamber, thus rendering it conductive and permitting a current flow through the air between two charged electrodes. This gives the sensing chamber an effective electrical conductance. When smoke particles enter the ionization area, they decrease the conductance of the air by attaching themselves to the ions, causing a reduction in mobility. When the conductance is less than a predetermined level, the detector responds.

Smoke detectors utilizing the photoelectric light-scattering principle are usually of the spot-type. They contain a light source and a photosensitive device so arranged that light rays do not normally fall onto the photosensitive device. When smoke particles enter the light path, light strikes the particles and is scattered onto the photosensitive device, causing the detector to respond.

Smoke detectors utilizing the photoelectric light-obscuration principle consist of a light source, which is projected onto a photosensitive device. Smoke particles between the light source and the photosensitive device reduce the light reaching the device causing the detector to respond.

Cloud chamber smoke detection is accomplished by drawing an air sample from the protected area into a high humidity chamber and lowering the chamber pressure to create an environment in which the resultant moisture in the air condenses on any smoke particles present, forming a cloud. The cloud density is measured by photoelectric principle. The density signal is processed and used to convey an alarm condition when it meets preset criteria.

Radiant energy detectors include flame detectors and spark/ember detectors. Radiant energy includes the electromagnetic radiation emitted as a by-product of the combustion reaction, which obeys the laws of optics. This includes radiation in the ultraviolet (0.1 to 0.35 μm), visible (0.36 to 0.75 μm), and infrared (0.76 to 220 μm) portions of the spectrum emitted by flames or glowing embers.

(NOTE: 1.0 micron = 1000 nanometers = 10,000 angstroms)

Ultraviolet flame detectors typically use a vacuum photodiode Geiger-Muller tube to detect the ultraviolet radiation that is produced by a flame. The photodiode allows a burst of current to flow for each ultraviolet photon that hits the active area of the tube. When the number of current bursts per unit time reaches a predetermined level, the detector initiates an alarm. Other types of flame detectors include:

1) Single wavelength infrared flame detector

2) Ultraviolet/Infrared (UV/IR) flame detector

3) Multiple wavelength infrared (IR/IR)

A spark/ember detector is a radiant energy fire detector that is designed to detect sparks or embers or both. These devices are normally intended to operate in dark environments and in the infrared part of the spectrum.

Fire-gas detectors respond to one or more of the gases produced by a fire. The operating principles of fire-gas detectors are semiconductor and catalytic element.

Sprinkler waterflow alarm-initiating devices indicate the flow of water in a building fire sprinkler system.

NFPA 72-1999 2-6.2

Initiation of a sprinkler waterflow alarm signal shall occur within 90 seconds of waterflow at the alarm-initiating device when waterflow that is equal to or greater than the flow from a single sprinkler of the smallest orifice installed in the system occurs. Movement of water due to waste, surges, or variable pressure shall not be indicated.

Supervisory signal-initiating devices include:

1) Control valve

2) Pressure

3) Water level

4) Water temperature

5) Room temperature

NFPA 72-1999 4-3.1.1

An average sound level greater than 105dBA shall require the use of a visible signal appliance in accordance with Section 4-4.

NFPA 72-1999 4-3.1.4

Where audible appliances are installed in mechanical equipment rooms, the average ambient sound level used for design guidance shall be at least 85dBA for all occupancies.

NFPA 72-1999 4-3.2.1

Audible signal appliances intended for use in the public mode shall have a sound pressure level of not less than 75 dBA at 10 ft nor more than 120 dBA at the minimum hearing distance from the audible appliance.

NFPA 72-1999 4-3.2.2

To ensure that audible public mode signals are clearly heard, they shall have a sound pressure level at least 15dBA above average ambient sound pressure level or 5 dBA above the maximum sound pressure level having a duration of at least 60 seconds, whichever is greater, measured 5 ft above the floor in the occupiable space.

NFPA 72-1999 4-3.3.1

Audible signals intended for operation in the private mode shall have a sound pressure level of not less than 45 dBA at 10 ft nor more than 120dBA at the minimum hearing distance from the audible appliance.

NFPA 72-1999 4-3.4

Audible appliances installed to notify occupants in sleeping areas, they shall have a sound level of at least 15 dBA above the average ambient sound pressure level or 5 dBA above the maximum sound level having a duration of at least 60 seconds or a sound level of at least 70dBA, whichever is greater, measured at the pillow level in the occupiable area.

NFPA 72-1999 4-3.5.1

Where ceiling heights allow, wall-mounted audible notification appliances shall have their tops above the finished floors of not less than 90 in. and below the finished ceiling of not less than 6 in.

NFPA 72-1999 4-3.5.2

Where combination audible/visible appliances are installed, the location of the installed appliance shall follow the requirements of 4-4.4 for visible notification appliances.

NFPA 72-1999 4-4.2

The flash rate of visible notification appliances shall not exceed two flashes per second (2 Hz) nor be less than one flash every second (1 Hz) throughout the listed voltage range of the visible notification appliance.

NFPA 72-1999 4-4.2.1

The maximum pulse duration of a visible notification appliance shall be 0.2 second with a maximum duty cycle of 40 percent. The pulse duration is defined as the time interval between initial and final points of 10 percent of maximum signal.

NFPA 72-1999 4-4.2.2

The light source color of a visible notification appliance shall be clear or nominal white and shall not exceed 1000 candela.

NFPA 72-1999 4-4.4

Wall-mounted visible notification appliances shall have their bottoms at heights above the finished floor of not less than 80 in. and no greater than 96 in.

NFPA 72-1999

Notes

31005 PERIODIC TESTS

REFERENCES:

NFPA 72

DESCRIPTION:

Know periodic equipment and circuit testing procedures, including frequency of tests and method of testing each component and circuit of a fire alarm system.

NFPA 72-1999 7-1.2

The owner is responsible for inspection, testing, maintenance, alterations, and additions to the fire alarm system.

NFPA 72-1999 7-1.2.1

Inspection, testing, and maintenance is permitted to be done by a person or organization other than the owner where conducted under written contract.

NFPA 72-1999 7-1.2.2

Service personnel shall be qualified and experienced in the inspection, testing, and maintenance of fire alarm systems. Examples of qualified personnel include individuals who are:

(a) Factory trained and certified.

(b) National Institute for Certification in Engineering Technologies fire alarm certified.

(c) International Municipal Signal Association fire alarm certified.

(d) Certified by a state or local authority.

(e) Trained and qualified personnel employed by an organization listed by a national testing laboratory for the servicing of fire alarm systems.

NFPA 72-1999 7-1.3.1

Before proceeding with testing, facilities that receive alarm, supervisory, or trouble signals, and all building occupants, shall be notified to prevent unnecessary response. At the conclusion of testing, those previously notified shall be notified that testing has been concluded.

NFPA 72-1999 7-1.4

Prior to system maintenance or testing, the system Record of Completion and information regarding the system and system alterations, including specifications, wiring diagrams, and floor plans, shall be provided by the owner to the service personnel.

NFPA 72-1999 7-1.6.2.1

Reacceptance testing shall be performed after system components are added or deleted; after any modification, repair, or adjustment to system hardware or wiring; or after any change to software. All components, circuits, systems operations, or site-specific software functions known to be affected by the change or identified by a means that indicates the system operational changes shall be 100 percent tested. In addition, 10 percent of initiating devices that are not directly affected by the change, up to a maximum of 50 devices, also shall be tested and proper system operation shall be verified. A revised Record of Completion in accordance with 1-6.2.1 shall be prepared to reflect any changes.

NFPA 72-1999 7-5.1

After successful completion of acceptance tests satisfactory to the authority having jurisdiction, a set of reproducible as-built installation drawings, operation and maintenance manuals, and a written sequence of operation shall be provided to the building owner or the owner's designated representative. It shall be the responsibility of the owner to maintain these records for the life of system and to keep them available for examination by any authority having jurisdiction. Paper or electronic media shall be permitted.

NFPA 72-1999 7-3.2.1

Smoke detector sensitivity shall be checked within 1 year after installation and every alternate year thereafter. After the second required calibration test, where sensitivity test indicate that the detector has remained within its listed and marked sensitivity range (or 4 percent obscuration light grey smoke, if not marked), the length of time between calibration tests shall be permitted to be extended to a maximum of 5 years. Where the frequency is extended, records of detector-caused nuisance alarms and subsequent trends of these alarms shall be maintained. In zones or in areas where nuisance alarms show any increase over the previous year, calibration tests shall be performed.

To ensure that each smoke detector is within its listed and marked sensitivity range, it shall be tested using either:

(a) A calibrated test method; or

(b) Manufacturer's calibrated sensitivity test instrument; or

(c) Listed control equipment arranged for the purpose; or

(d) A smoke detector/control unit arrangement whereby the detector causes a signal at the control unit where its sensitivity is outside its acceptable sensitivity range; or

(e) Other calibrated sensitivity test method acceptable to the authority having jurisdiction.

Detectors determined to have a sensitivity outside the listed and marked sensitivity range shall be cleaned and recalibrated or replaced. This requirement shall not apply to single station detectors referenced in 7-3.3 and Table 7-2.2.

Smoke detector sensitivity shall not be tested or measured using any device that administers an unmeasured concentration of smoke or other aerosol into the detector.

Refer to NFPA 72-1999 7-2.2 for Test Methods and NFPA 72-1999 7-3.1 & 7-3.2 for Testing Frequencies.

Notes

31006 BASIC ELECTRICITY

DESCRIPTION:

Understand DC circuits, use of Ohm's Law, series and parallel circuits, resistance of wires, voltage drop calculations, use of VOM (volt-ohm-milliammeter), and AC circuits.

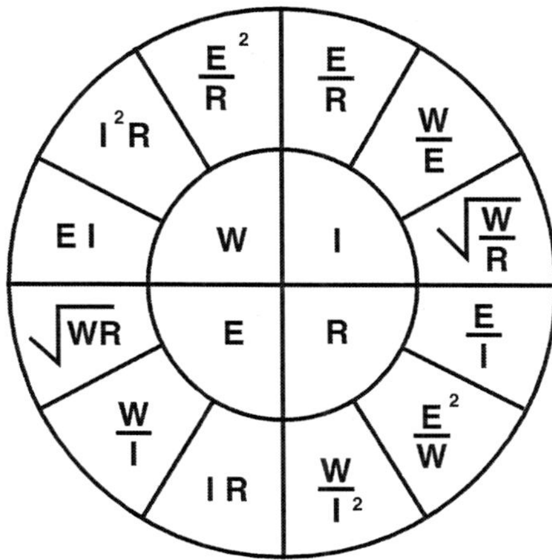

Ohm's Law Wheel

The four basic units of electricity:

- Electromotive force or voltage as measured in Volts

- Intensity or current as measured in Amperes

- **R**esistance as measured in Ohms

- Power as measured in **W**atts

Electricity is the flow of electrons.

One method to illustrate the behavior of electricity is to compare it to the way water behaves. Water in a tank exerts pressure on a valve and pipe for example, which is measured in pounds per square inch. As long as the valve remains closed, water is not allowed to flow and nothing happens, however, there exists the potential to perform work. Work will be performed when the valve is opened and water is allowed to flow through the pipe.

VOLTAGE

A battery can be compared to the tank of water. A battery has the potential to perform work, however, work will not be performed until the switch is closed and a circuit is formed allowing electrons to flow. When considering the battery, an electrical circuit represents the valve and pipe. Pressure in pounds per square inch is used to describe the potential energy of the water tank. In electricity, potential energy is described by voltage. Voltage can be thought of as electrical pressure, and is measured in volts. The battery is much like an electrical storage tank.

1 volt equals the potential to flow one ampere through 1 ohm

CURRENT

When the water tank valve is opened, water will flow at a measurable rate in gallons per minute, for example. The water flow rate is determined by water pressure and the diameter of the valve. Electricity will flow from the battery when a complete circuit exists. Just like the water tank, the rate at which electrons flow from the battery is dependent upon the battery voltage or pressure and the diameter of the wire (pipe). The rate at which electrons flow from the battery is measured in amperes.

1 Coulomb equals 6,250,000,000,000,000,000 electrons

1 amp equals 1 Coulomb per second

RESISTANCE

In the water tank, resistance to water flow can be caused by partially shutting the valve. Partially shutting the valve will hinder the flow of water. If the electrical circuit attached to a battery is partially closed by using a smaller conductor for example, heat will be generated which will hinder the flow of electrons. Electrical resistance is measured in Ohms.

1 Ohm will allow 1 Volt to push 1 Ampere of current

POWER

The amount of power produced by flowing water from a tank through a valve and pipe can be determined by the pressure of the water times the flow rate allowed by the valve and pipe and is measured in foot-pounds per second or horsepower. The amount of power produced by a circuit connected to a battery can be determined by the voltage of the battery times the current allowed by the circuit and is measured in Watts.

1 Volt pushing 1 Ampere through 1 Ohm will produce 1 Watt

RESISTORS

Resistors come in a variety of sizes and values. Because of this and because most resistors are small, a color code of banding has been developed to make it easier to identify resistor value. A resistor will ordinarily have 4 color bands. The first three indicate resistance and the fourth describes the tolerance value.

<u>RESISTOR COLOR CODE</u>

Black	=	0
Brown	=	1
Red	=	2
Orange	=	3
Yellow	=	4
Green	=	5
Blue	=	6
Violet	=	7
Grey	=	8
White	=	9

Tolerance

Gold	=	5%
Silver	=	10%
No color	=	20%

OHM'S LAW

George Ohm observed a relationship between Voltage, Current, and Resistance, which is known today as Ohm's law. When 1 Volt is connected to 1 Ohm, 1 Ampere of electricity will flow. In other words:

$$E = IR$$

E (Voltage) is equal to I (Current) times R (Resistance)

or;

$$I = E/R$$

I (Current) is equal to E (Voltage) divided by R (Resistance)

or;

$$R = E/I$$

R (Resistance) is equal to E (Voltage) divided by I (Current)

and;

$$W = IE$$

W (Power) is equal to I (Current) times E (Voltage)

or;

$$I = W/E$$

I (Current) is equal to W (Power) divided by E (Voltage)

or;

$$E = W/I$$

E (Voltage) is equal to W (Power) divided by I (Current)

Another way to illustrate Ohms Law is to use a triangle. In the triangle below you will find three units of Ohms Law. If you want to solve for E (Voltage), cover the E and you find that you should multiply I (Current) times R (Resistance).

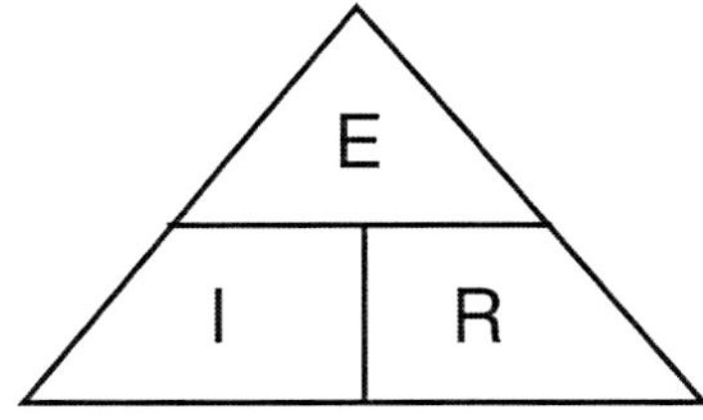

RULES TO FOLLOW WHEN USING OHM'S LAW

RULE #1. Whenever possible solve parallel portions of a circuit first then solve the series portion.

RULE #2. Voltage adds across a series circuit and remains constant throughout a parallel circuit.

RULE #3. Current adds across a parallel circuit and remains constant throughout a series circuit.

PARALLEL RESISTANCE FORMULAS

2 Resistors in parallel

"Product over the Sum"

$$R_T = \frac{R_1 \times R_2}{R_1 + R_2}$$

2 or more resistors in parallel

"The reciprocal of the sum of the reciprocals"

$$R_T = \frac{1}{\frac{1}{R_1} + \frac{1}{R_2} + \frac{1}{R_3}}$$

SERIES RESISTANCE FORMULA

$$R_T = R_1 + R_2 + R_3$$

EXAMPLES

The circuit below includes three resistors; R1 = 50 ohms, R2 = 40 ohms, and R3 = 30 ohms. R2 is in parallel with R3 and R1 is in series with the circuit represented by R2 and R3. Solve the parallel portion first by using "The Product over the Sum" method. Multiply R2 times R3 and divide the result by the sum of R2 and R3. 40 ohms times 30 ohms equals 1200 ohms divided by 70 ohms (40 ohms plus 30 ohms) equals 17.14 ohms. The next step is to solve the resulting series circuit of R1 and the combination of R2 and R3; 50 ohms plus 17.14 ohms or 67.14 ohms.

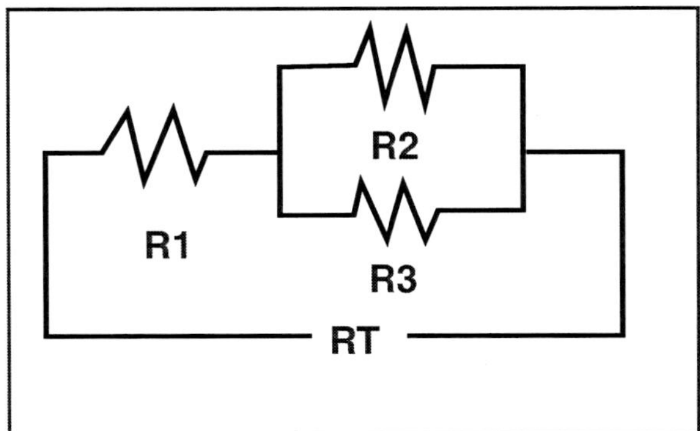

USING A CALCULATOR TO SOLVE RESISTORS IN PARALLEL

Here's an easy method to determine the total resistance of 2 or more resistors in parallel.

1. Turn your calculator on and clear the memory.

2. For each resistor value press;

   "1", "divide button", "value for R1", "M+"

   "1", "divide button", "value for R2", "M+"

   "1", "divide button", "value for R3", "M+"

   and so on for each resistor in parallel.

3. Once all of the resistor values have been entered, press;

   "1", "divide button", "MR", "="

PARALLEL VOLTAGE FORMULA

$$E_T = E_1 = E_2 = E_3 = E_4$$

PARALLEL CURRENT FORMULA

$$I_T = I_1 + I_2 + I_3 + I_4$$

SERIES VOLTAGE FORMULA

$$E_T = E_1 + E_2 + E_3 + E_4$$

SERIES CURRENT FORMULA

$$I_T = I_1 = I_2 = I_3 = I_4$$

CAPACITORS

In general, a capacitor consists of two conducting surfaces separated by an insulating material. The conducting surfaces are called plates and the insulating material is called the dielectric. The dielectric material can be any good insulator, including air. The structure and composition will vary from one type of capacitor to the next, but the purpose of all capacitors to store electric energy supplied by an external source.

When voltage is applied to a capacitor, electrons collect on the plates. Electrons will continue to collect on the plates until the capacitor plates are at the same potential as the power supplied. When the circuit is opened and the power supplied to the capacitor is disconnected, and there is no circuit for the electrons to move through, the capacitor is considered to be charged. The electrons stay on the plates at whatever the polarity they were applied. The capacitor will hold this charge until a suitable discharge path is provided. In order for discharge to take place, there must be a complete circuit between the plates. When the path is provided, the electrons move from the most negative plate to the most positive plate until the difference in potential is zero at which point the capacitor is discharged. There is no difference of potential between the plates, and current flow stops.

When the applied voltage increases, the capacitor will charge to a higher value matching the applied voltage. When the applied voltage decreases, the capacitor will discharge thus maintaining voltage for a period of time.

The capacitor's ability to store and discharge opposes changes in circuit voltage. This ability is defined as Capacitance and is measured in Farads, named for Michael Faraday.

1 Farad is equal to 1 Coulomb of charge per Volt of pressure

When a capacitor is connected across a battery or other power source, work is performed to charge the capacitor and store a charge. When the capacitor discharges, the stored energy is returned to the circuit. The charge of the capacitor is a measure of the energy it can store, which corresponds to the quantity of electrons transferred divided by the amount of voltage used to store it.

Since one farad represents one coulomb per volt, the expression may be written as an equation:

$$C = Q/E$$

C = Capacitance in farads
Q = Charge in coulombs
E = voltage in volts

Like ohm's law, the capacitor charge expression contains three quantities, and may be manipulated to determine an unknown quantity when the other two are known:

$$C = Q/E \text{ or;}$$

$$Q = CE \text{ or;}$$

$$E = Q/C \text{ or;}$$

CAPACITOR VOLTAGE RATING

When too much voltage is applied to a capacitor the dielectric molecules will distort until electrons are pulled from the atoms. These free electrons will flow through the dielectric from one plate to the other, changing the characteristics of the capacitor. In some cases the current resulting from this condition will create enough heat to burn a hole through the dielectric. The voltage at which this condition occurs is called the **Breakdown Voltage**. The ability of a dielectric material to withstand voltage breakdown is known as **Dielectric Strength**. To safeguard against this type of breakdown, capacitors must be operated at voltages below their dielectric breakdown voltage. The maximum voltage that can be continuously to a capacitor without damage is called the **Working Voltage**. Working voltage is determined by dielectric strength and dielectric material thickness.

There is no such thing as a perfect insulator, therefore, after a capacitor is charged, leakage current through the dielectric will cause the charge to leak off in time. Some capacitors will remain charged for several minutes while others will retain a charge for several hours or months, depending upon the amount of leakage current.

The dielectric characteristics change due to aging, heat, and humidity. Leakage and the dielectric constant also tend to change over time. All of these factors will change the value of capacitance.

Capacitors can be combined in series or parallel to produce different values of capacitance:

CAPACITORS IN PARALLEL
Total capacitance is equal to the sum of the Individual capacitance values in a parallel circuit.

$$C_T = C_1 + C_2 + C_3 + C_4$$

CAPACITORS IN SERIES
Total capacitance in a series circuit is the reciprocal of the sum of the reciprocals.

$$C_T = \cfrac{1}{\dfrac{1}{C_1} + \dfrac{1}{C_2} + \dfrac{1}{C_3}}$$

SAMPLE PROBLEMS IN OHM'S LAW

SERIES CIRCUIT

1.

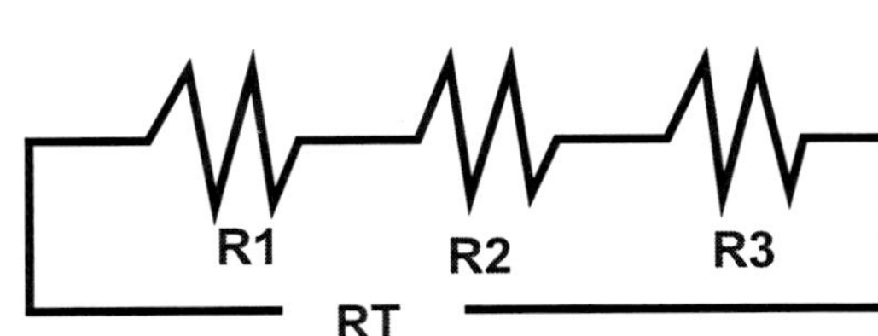

R_1 = 20 ohms R_2 = 25 ohms R_3 = 15 ohms

R_T = _______

I_1 = .4 amp I_2 = .4 amp I_3 = .4 amp

I_T = _______

E_1 = 8 volts E_2 = 10 volts E_3 = 6 volts

E_T = _______

PARALLEL CIRCUIT

2.

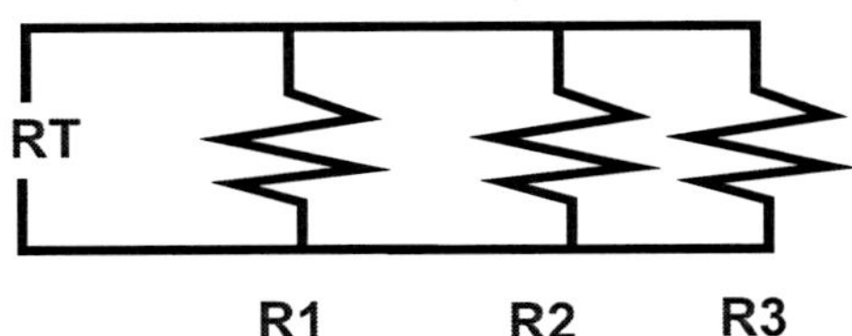

R_1 = 60 ohms R_2 = 80 ohms R_3 = 120 ohms

R_T = _______

I_1 = 2 amps I_2 = 1.5 amps I_3 = 1 amp

I_T = _______

E_1 = 120 volts E_2 = 120 volts E_3 = 120 volts

E_T = _______

3. # COMBINATION CIRCUIT

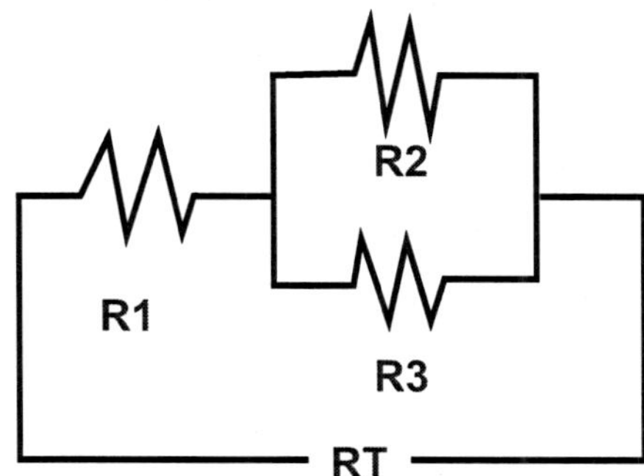

R_1 = 10 ohms R_2 = 20 ohms R_3 = 30 ohms

R_T = ______

I_1 = ______ I_2 = ______ I_3 = ______

I_T = ______

E_1 = ______ E_2 = ______ E_3 = ______

E_T = 110 volts

4.

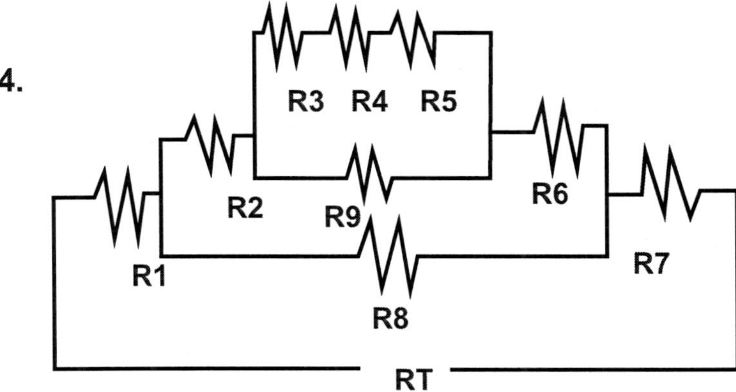

R_1 = 15 R_6 = 25

R_2 = 35 R_7 = 10

R_3 = 50 R_8 = 300

R_4 = 40 R_9 = 60

R_5 = 30 R_t = _________

VOLTAGE DROP

As electrons flow through a circuit, energy is depleted as the load performs work and through resistance in the form of heat. The power source provides additional electron flow as needed. The portion of energy consumed as a result of circuit resistance can be predicted by calculating voltage drop based upon circuit resistance.

NFPA 70 *National Electrical Code* provides two guidelines regarding permissible voltage drop:

1. The percent of drop shall not exceed 5% from the source through the last overcurrent protection device to the load.

2. The circuit from the last overcurrent device (panel fuse) and back again, exclusive of the load will not be more than 3% of the total source current.

Following these guidelines, the total voltage drop allowed in a 12VDC circuit is 3% of 12VDC or .36VDC. The total voltage drop allowed in a 24VDC fire alarm control panel is 3% of 24VDC or .72VDC. Consideration of circuit resistance based upon the length and size of conductors is required to meet these guidelines.

CALCULATING VOLTAGE DROP

Calculating voltage drops in a notification appliance circuit is rather straightforward once we establish which mathematical formulae to use. As is the case with any problem to be solved, we must first determine what we are looking for and in what units. In this case, we want to determine what the voltage drop will be in a circuit such that we can supply notification appliances with the voltage and current recommended by the manufacturer. Volts are what we are looking for and shall use Vd to designate the units. The voltage drop can be stated arithmetically as:

Voltage Drop = Voltage (at the power supply) − Voltage (at the notification appliance)

$$Vd = Vs - Vn$$

For example, we put an existing fire alarm system into alarm and use a voltage meter to determine that the fire alarm control panel is providing 28 VDC to the notification appliance circuit. Using the same voltage meter, we determine that there is 25 VDC at the notification appliance while it is in alarm:

$$Vd = 28VDC - 25VDC = 3VDC$$

Now that we know what it is we are looking for, lets consider the litany of algebraic formula available to help us in our quest. Before choosing a formula, let's review what we know.

We will know the supply voltage provided by the fire alarm control panel or notification appliance power supply so lets use E to designate supply voltage.

We can determine how much current each of our notification appliances will require to operate properly by reading the nameplate data or instruction manual, so lets use I to designate current.

We can figure out how much circuit resistance we will have to overcome in part by measuring the wire runs from the supply voltage to the notification appliance. To determine circuit resistance due to copper conductors, we can use the chart in NFPA 70-1999, Table 8. Once we know the wire length in feet and the conductor resistance per foot in ohms we can use Ohm's Law to determine circuit resistance:

$$R = E/I$$

For example, we have installed 1000 feet of 16AWG 2-conductor solid uncoated copper wire and used it to connect a horn-strobe to a fire alarm control panel. NFPA 70-1999 Table 8 tells us that at 75 degrees Centigrade 16AWG solid uncoated copper wire exhibits 4.89 ohm of resistance per 1000 feet. Restated, .00489 ohm of resistance per foot. 1000 feet (times two for 2-conductor cable) times .00489 ohm/ft = 9.78 ohm. In this example, R = 9.78 ohm

To review, we want to solve for Vd and we know R, .00978 ohm. We know from the nameplate data that the notification appliance requires .100 amp.

$$Vd = IR$$

$$Vd = (.100 \text{ amp}) \times (9.78 \text{ ohm}) = .978 \text{ volts}$$

If the supply voltage is 28 volts, then .978 volts is about a 3% drop.

VOLTAGE DROP FORMULAS

NOTATIONS

I - amps   R = resistance   VD = voltage drop

VD permitted = 3% of source voltage

CM = size of conductor in circular mils *(NFPA Table 8)*

D = distance of the circuit one way

PL = power loss (wasted electricity) due to conductor resistance

K = resistance of a circular mil-foot of wire *(Table 8)*

Ⓚ = approximate K; use 12.9 for copper; 21.2 for aluminum

$$\textbf{Power Loss} = VD \times I \qquad \textbf{Exact K} = \frac{R \times CM}{1000'}$$

$$\textbf{Voltage Drop} = VD = \frac{2 \times K \times D \times I}{CM} \quad \text{or} \quad \textbf{Voltage Drop} = VD = I \times R$$

$$\textbf{Wire Size} = CM = \frac{2 \times Ⓚ \times D \times I}{VD \text{ permitted}}$$

$$\textbf{Distance} = D = \frac{CM \times VD \text{ permitted}}{2 \times K \times I}$$

$$\textbf{Load} = I = \frac{CM \times VD \text{ permitted}}{2 \times K \times D}$$

Using this information we can solve most problems regarding voltage drop.

VOLTAGE DROP PROBLEMS

1. What is the total resistance of 40 feet of #18 AWG uncoated copper wire?

2. What is the maximum distance allowed between a 12-volt source and load requiring .75 amp using #16AWG coated copper wire?

3. What is the minimum sized copper conductor required for a 1.5 amp load, located 75 feet from a 24-volt supply?

4. A #16AWG stranded uncoated copper conductor has a total resistance of 0.43 ohms. What is the approximate length of this conductor?

5. What is the voltage drop in a branch circuit 40 feet long with a supply voltage of 6 volts connected to a 50 ohm load with #14AWG copper conductor?

6. What is the total resistance of two, 85 foot, #16AWG copper conductors connected in parallel?

7. What is the maximum load in amps allowed by NFPA 70 for a 65-foot long branch circuit using #18AWG coated copper wire connected to a 12-volt source?

8. What is the minimum sized copper conductor allowed for a 50- foot long branch circuit connecting a 6-volt source to a .25 amp load.

9. What is the approximate length of a #18AWG stranded uncoated copper branch circuit with a total resistance of 1.05 ohm.

10. What is the voltage drop in a 56-foot #14AWG stranded copper branch circuit connecting a 24-volt source to a .36 amp load?

OHM'S LAW ANSWERS

1. Remember the 3 rules for Ohm's Law. In a series circuit total resistance and total voltage are the sum of the individual values. Therefore **R_t = 60 ohms** and **E_T = 24 volts**. In a series circuit current is constant, therefore **I_T = .4 amp**.

2. Remember again the 3 rules of Ohm's Law. In a parallel circuit voltage is constant, therefore **E_T = 120 volts**. Total current in a parallel circuit is the sum of the individual values therefore **I_t = 4.5amps**. Use your calculator to solve for **R_T:**

    Clear Memory

    1 divided by 60, then press the memory add button

    1 divided by 80, then press the memory add button

    1 divided by 120, then press the memory add button

    The answer is found by pressing 1 divided by the memory recall button = **26.666737 ohms**

3. The first step to solve this problem is to solve for R_A. R_A is the value of resistance represented by R_2 and R_3 in parallel. Using your calculator:

    Clear memory

    1 divided by 20, then press the memory add button

    1 divided by 30, then press the memory add button

    1 divided by the memory recall button = **12 ohms = R_A**

Now use the value of R_A to solve the series circuit which includes R_A and R1 by adding the two values together **= 22 ohms**.

Next solve for I_T = E_T/R_T = 110/22 or **I_T = 5amps**.

Because current remains the same in a series circuit, **I_A** and **I_1** are both **5amps**.

Next solve for E_1 = $I_1 \times R_1$ = 5 x 10 **= 50 volts**

And E_T - E_1 = E_A or 110 - 50 = **E_A = 60 volts**

Next, because voltage in a parallel circuit remains constant or E_A = E_2 = E_3, we know that **E_2 = 60 volts** and **E_3 = 60 volts**

Finally, I_2 = E_2/R_2 = 60/20 = **3amps**

And I_3 = E_3/R_3 = 60/30 = **2amps**

4. In order to solve this remember to solve for the parallel portion of circuits first:

$R_A = R_3 + R_4 + R_5 =$ **120**

Now solve the parallel equivalent of R_A and R_9;

$$R_B = \frac{120 \times 60}{120 + 60} = 40$$

Next determine the value of R_B, R_2, and R_6 in series; **R_C = 100**

Next determine the value of R_C in parallel with R_8 to get **R_D = 75**

Now solve for the value of $R_1 + R_D + R_7 =$ **R_T = 100**

VOLTAGE DROP ANSWERS

1. From NFPA 70 Table 8 we get 7.77 ohms / 1000 ft. or;

    .00777 ohm / ft.

    40 ft. X .00777 ohm / ft. = .3108 ohm.

2. First solve for exact K. $$K = \frac{R / 1000 \text{ ft.} \times CM}{1000}$$

    #16AWG coated copper wire = 5.08 ohm / 1000 ft.

    #16AWG coated copper wire = 2,580 CM

    K = 5.08 ohm x 2580 CM = 13.1064

    Now solve for maximum Distance. $$D = \frac{CM \times VD \text{ permitted}}{2 \times K \times I}$$

    $$\frac{2580 \text{ CM} \times 0.36 \text{ VD permitted}}{2 \times 13.1064 \times 0.75 \text{ amp}} = \textbf{47.2 feet}$$

3. $CM = \dfrac{2 \times \textcircled{K} \times D \times I}{VD\ permitted}$

$CM = \dfrac{2 \times 12.9 \times 75\ feet \times 1.5\ amp}{.72\ volt}$

$CM = 4031$

From NFPA 70 Table 8 4,110 CM = **#14AWG**

4. From NFPA 70 Table 8 #16AWG = 4.99 ohm/1000 ft. or; 0.00499 ohm / ft.

$\dfrac{.43\ ohm}{.00499\ ohm\ /\ ft.}$ = **86.1 feet**

5. VD = I x R

1) I = E/R = 6 volts / 50 ohms = .12 amp

2) From NFPA 70 Table 8 #14AWG = 3.07 ohm/1000 ft. or; .00307 ohm / ft.

Therefore R = 80 ft. x .00307 ohm / ft. = .2456 ohm.

3) VD = .12amp x .2456ohm = **0.029472 volt**

6. NFPA 70 Table 8 #16AWG copper = 4.89 ohm / 1000 ft.

.00489 ohm / ft. x 85 feet = .416 ohm

$\dfrac{.416\ ohm}{2}$ = **0.208 ohm**

7. $I = \dfrac{CM \times VD \text{ permitted}}{2 \times K \times D}$

Exact $K = \dfrac{R/1000 \text{ ft.} \times CM}{1000}$

$K = \dfrac{8.08 \text{ ohm} \times 1620 \text{ CM}}{1000} = 13.0896$

$I = \dfrac{1620 \text{ CM} \times .36 \text{ volt}}{2 \times 13.0896 \times 65 \text{ ft.}} = \textbf{0.34 amp}$

8. $CM = \dfrac{2 \times Ⓚ \times D \times I}{VD \text{ permitted}}$

$CM = \dfrac{2 \times 12.9 \times 50 \text{ ft.} \times .25 \text{ amp}}{}$

$CM = 1,791 \text{ CM}$

NFPA 70 Table 8 #16AWG copper conductor = **2,580 CM**

9. NFPA 70 Table 8 #18AWG = 7.95 ohm / 1000 ft. or;

0.00795 ohm per ft.

$\dfrac{1.05 \text{ohm}}{.00795 \text{ohm} / \text{ft.}}$ $=$ $\dfrac{132 \text{ feet of wire}}{2 \text{ conductors}} =$ **66 ft.**

10. VD = I x R

I = 0.36 amp

From NFPA 70 table 8 #14AWG = 3.14 ohm / 1000 ft.

.00314 ohm / ft. x 56 ft. x 2 conductors = 0.35168 ohm

VD = 0.36 amp x 0.35168 ohm = **0.1266 volt**

31007 BASIC WORKING DRAWINGS

REFERENCES:

NFPA 72

NFPA 170

Drafting Texts

DESCRIPTION:

Understand the basic requirements of fire alarm system working drawings. Demonstrate good drafting techniques. Prepare simple layouts by use of good standard practice with proper line widths and letters. Produce work that meets standards of the employer and is of good standard practice in the industry. Calculate quantity and type of automatic fire detectors required and prepare job material lists.

Glossary of Terms

Floor Plans are views of a building drawn to scale. Rooms, hallways, cabinets, etc. are drawn in the correct relationship to each other. A legend is used to identify devices using symbols. NFPA 170 includes the devices found in fire alarm systems.

Lines. The lines used to make a blueprint are typically of one of three widths; thin (.010"), medium (.015"), and thick (.020"). Types of lines include:

Border line. A thick line used to identify the edge or border of a blueprint.

Object line. A thick line used to depict an object or structure.

Dimension line. A thin line used to define a dimension terminating with an arrow at each end. Dimension lines should never pass through an object.

Center line. A thin line of alternating long and short dashes which always ends with a long dash. Centerlines are used to represent the axis of symmetrical objects and to indicate the center. Centerlines are used to indicate the position of doors and windows.

Extension line. A thin line disconnected from an object used to indicate the end of a dimension line.

Cutting plane line. A thick broken line used to illustrate where a cross section drawing exists usually identified with an 'A' or 'AA', for example.

Section line. Thin evenly spaced crosshatched lines used to indicate a cross-section.

Hidden line. Broken (dotted) medium width line used to represent either that something exists above a cutting plane line or that something is hidden.

Break line. A thick jagged line drawn freehand to indicate a short break. A thin straight line interrupted by zigzags indicates a long break.

Symbols. A litany of small diagrams or icons used instead of words to identify objects, materials, and product types on a drawing.

An orthographic, right angle, or projection drawing include projector lines, which are parallel to each other and perpendicular to the plan of projection.

An oblique projection include projector lines which are parallel to each other and oblique to the plane of projection. (Oblique is defined as slanting or sloping).

When the receding lines are a true length, that is when the projector lines form an angle of 45° with the plane of projection, the oblique drawing is called a cavalier projection.

Identify fire alarm symbols as represented in NFPA 170.

Understand the differences in line types used in drawings.

<table>
<tr><td>NFPA 72-1999 1-6.2.2</td><td>Every system shall include the following documentation, which shall be delivered to the owner or the owner's representative upon final acceptance of the system.</td></tr>
</table>

    (a) An owner's manual and installation instructions covering all system equipment; and

    (b) Record drawings.

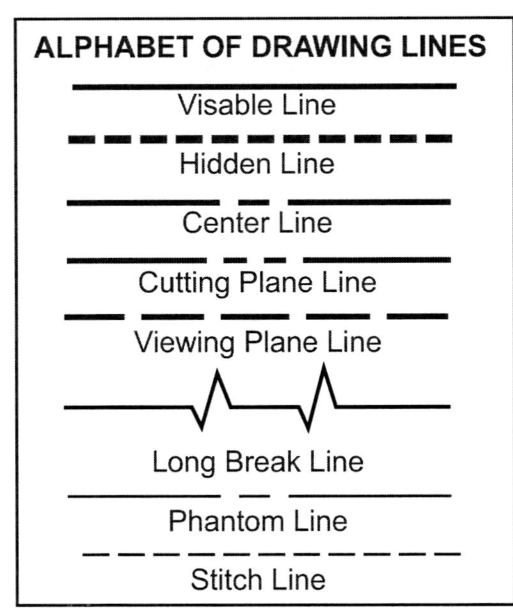

Electrical Reference Symbols

ELECTRICAL ABBREVIATIONS
(Apply only when adjacent to an
electrical symbol.)

Central Switch Panel	CSP
Dimmer Control Panel	DCP
Dust Tight	DT
Emergency Switch Panel	ESP
Empty	MT
Explosion Proof	EP
Grounded	G
Night Light	NL
Pull Chain	PC
Rain Tight	RT
Recessed	R
Transfer	XFER
Transformer	XFRMR
Vapor Tight	VT
Water Tight	WT
Weather Proof	WP

ELECTRICAL SYMBOLS

Switch Outlets

Single-Pole Switch	S
Double-Pole Switch	S_2
Three-Way Switch	S_3
Four-Way Switch	S_4
Key-Operated Switch	S_K

Switch and Fusestat Holder	S_{FH}
Switch and Pilot Lamp	S_P
Fan Switch	S_F
Switch for Low-Voltage Switching System	S_L
Master Switch for Low-Voltage Switching System	S_{LM}
Switch and Single Receptacle	S
Switch and Duplex Receptacle	S
Door Switch	S_D
Time Switch	S_T
Momentary Contact Switch	S_{MC}
Ceiling Pull Switch	(S)
"Hand-Off-Auto" Control Switch	HOA
Multi-Speed Control Switch	M
Push Button	•

Receptacle Outlets

Where weather proof, explosion proof, or other
specific types of devices are to be required, use the
upper-case subscript letters. For example, weather
proof single or duplex receptacles would have the
uppercase WP subscript letters noted alongside of
the symbol. All outlets should be grounded.

Single Receptacle Outlet	
Duplex Receptacle Outlet	
Triplex Receptacle Outlet	
Quadruplex Receptacle Outlet	

Duplex Receptacle Outlet
(Split Wired)

Triplex Receptacle Outlet
(Split Wired)

250-V Receptacle Single Phase.
Use Subscript Letter to Indicate
Function (DW—Dishwasher;
RA—Range, CD—Clothes Dryer) or
Numeral (with explanation in
symbol schedule).

250-V Receptacle Three Phase

Clock Receptacle

Fan Receptacle

Floor Single Receptacle Outlet

Floor Duplex Receptacle Outlet

Floor Special-Purpose Outlet

Floor Telephone Outlet (Public)

Floor Telephone Outlet (Private)

Example of the use of several floor outlet symbols
to identify a 2, 3, or more gang flow outlet

Underfloor duct and junction box
for triple, double, or single
duct system as indicated by
the number of parallel lines.

Example of use of various symbols to identify location of different types of outlets or connections for
underfloor duct or cellular floor systems

Cellular floor
Header duct

* Use numeral keyed to explanation in drawing list of symbols to indicate usage

Circuiting

Wiring Exposed (not in conduit)

Wiring Concealed in Ceiling
or Wall

Wiring Concealed in Floor

Wiring Existing *

Wiring Turned Up

Wiring Turned Down

Branch Circuit Home Run to
Panel Board.

Number of arrows indicates number of circuits.
(A number of each arrow may be used to identify
circuit number.)†

Bus Ducts and Wireways

Trolley Duct ‡

Busway (Service, Feeder, or
Plug-in)‡

Cable Trough Ladder or
Channel‡

Wireway ‡

**Panelboards, Switchboards,
and Related Equipment**

Flush Mounted Panelboard
and Cabinet‡

Surface Mounted Panelboard
and Cabinet‡

Switchboard, Power Control
Center, Unit Substations
(Should be drawn to scale.)‡

Flush-Mounted Terminal Cabinet
(In small scale drawings the
TC may be indicated alongside
the symbol.)‡

Surface-Mounted Terminal Cabinet
(In small scale drawings the
TC may be indicated alongside
the symbol.)‡

Pull Box (Identify in relation to Wiring System Section and Size.)

Motor or Other Power Controller (may be a starter or contactor)‡

Externally Operated disconnection Switch‡

Combination Controller and Disconnection Means‡

Power Equipment

Electric Motor (hp as indicated)

Power Transformer

Pothead (Cable Termination)

Circuit Element (e.g., Circuit Breaker)

Circuit Breaker

Fusible Element

Single-Throw Knife Switch

Double-Throw Knife Switch

Ground

Battery

Contactor

Photoelectric Cell

Voltage Cycles, Phase   Ex: 480/60/3

Relay

Equipment Connection (as noted)

*Note Use heavy-weight line to identify service and feeders. Indicate empty conduit by notation CO (conduit only)
†Note Any circuit without further identification indicates two-wire circuit. For a greater number of wires, indicate with cross lines, e.g.

3 wires   4 wires

Neutral wire may be shown longer. Unless indicated otherwise, the wire size of the circuit is the minimum size required by the specification. Identify different functions of wiring system, e.g., signalling system by notation or other means
‡Identify by notation or schedule

Remote Control Stations for Motors or Other Equipment

Push Button Station

Float Switch (Mechanical)

Limit Switch (Mechanical)

Pneumatic Switch (Mechanical)

Electric Eye (Beam Source)

Electric Eye (Relay)

Temperature Control Relay Connection (3 Denotes Quantity.)

Solenoid Control Valve Connection

Pressure Switch Connection

Aquastat Connection

Vacuum Switch Connection

Gas Solenoid Valve Connection

Flow Switch Connection

Timer Connection

Limit Switch Connection

Lighting

	Ceiling	Wall
Surface or Pendant Incandescent Fixture (PC = pull chain)	TYPE WATTS	SWITCH PC CIRCUIT
Surface or Pendant Exit Light		
Blanked Outlet		
Junction Box		

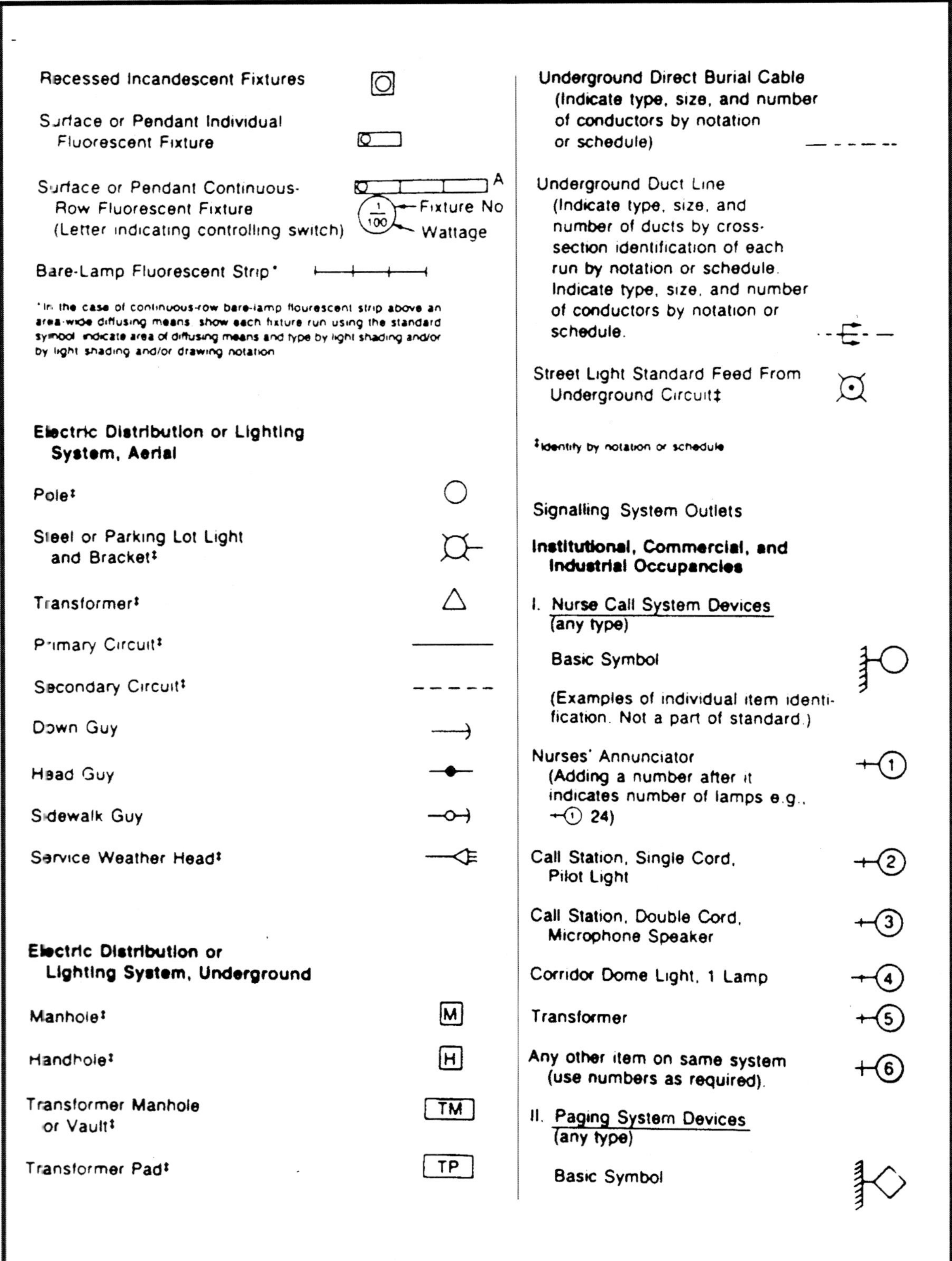

Recessed Incandescent Fixtures

Surface or Pendant Individual Fluorescent Fixture

Surface or Pendant Continuous-Row Fluorescent Fixture (Letter indicating controlling switch)

Fixture No
Wattage

Bare-Lamp Fluorescent Strip*

*In the case of continuous-row bare-lamp flourescent strip above an area-wide diffusing means, show each fixture run using the standard symbol. Indicate area of diffusing means and type by light shading and/or by light shading and/or drawing notation

Electric Distribution or Lighting System, Aerial

Pole‡

Steel or Parking Lot Light and Bracket‡

Transformer‡

Primary Circuit‡

Secondary Circuit‡

Down Guy

Head Guy

Sidewalk Guy

Service Weather Head‡

Electric Distribution or Lighting System, Underground

Manhole‡

Handhole‡

Transformer Manhole or Vault‡

Transformer Pad‡

Underground Direct Burial Cable (Indicate type, size, and number of conductors by notation or schedule)

Underground Duct Line (Indicate type, size, and number of ducts by cross-section identification of each run by notation or schedule. Indicate type, size, and number of conductors by notation or schedule.

Street Light Standard Feed From Underground Circuit‡

‡Identify by notation or schedule

Signalling System Outlets

Institutional, Commercial, and Industrial Occupancies

I. Nurse Call System Devices (any type)

Basic Symbol

(Examples of individual item identification. Not a part of standard.)

Nurses' Annunciator (Adding a number after it indicates number of lamps e.g., 24)

Call Station, Single Cord, Pilot Light

Call Station, Double Cord, Microphone Speaker

Corridor Dome Light, 1 Lamp

Transformer

Any other item on same system (use numbers as required).

II. Paging System Devices (any type)

Basic Symbol

(Examples of individual item identification Not a part of standard)

Keyboard

Flush Annunciator

Two-Face Annunciator

Any other item on same system · use numbers as required

III Fire Alarm System Devices
(any type) including Smoke and Sprinkler Alarm Devices

Basic Symbol

(Examples of individual item identification. Not a part of standard)

Control Panel

Station

10" Gong

Presignal Chime

Any other item on same system
(use numbers as required)

IV Staff Register System Devices
(any type)

Basic Symbol

(Examples of individual item identification. Not a part of standard)

Phone Operators' Register

Entrance Register (flush)

Staff Room Register

Transformer

Any other item on same system
(use number as required).

V Electric Clock System Devices
(any type)

Basic Symbol

(Examples of individual item identification Not a part of standard)

Master Clock

12" Secondary (flush)

12" Double Dial (Wall Mounted)

18" Skeleton Dial

Any other item on same system
(use numbers as required)

VI Public Telephone System Devices

Basic Symbol

(Examples of individual item identification. Not a part of standard)

Switchboard

Desk Phone

Any other item on same system
(use numbers required)

VII Private Telephone System Devices
(any type)

Basic Symbol

(Examples of individual item identification. Not a part of standard)

Switchboard

Wall Phone

Any other item on same system
(use numbers as required).

VIII. System Devices
(any type)

Basic Symbol

(Examples of individual item identification. Not a part of standard)

Notes

31008 BASIC MATHEMATICS

DESCRIPTION:

Solve mathematical problems requiring simple addition, subtraction, multiplication, division, and raising numbers to exponential powers. Round to the correct number of significant figures, calculate percentages, read graphs, and use simple geometric definitions and formulas.

Percentages are found by dividing the difference or change in a value by the original value. Then multiply this answer by 100 to convert to a percentage. The resulting percentage will be a plus or an increase for a positive change, or a negative or decrease for a negative change.

What percentage is 33 of 132?

Divide 33 by 132 = .25

Multiply .25 x 100 = 25%

What percentage is 42 of 16?

Divide 42 by 16 and multiply by 100 = 262.5%

Gasoline increased in price from $1.00 per gallon to $1.25 per gallon, what is the percentage increase in price?

1.25 - 1.00 = 25

25 x 100 = 25%

Gasoline decreased in price from $1.30 per gallon to $1.15 per gallon, what is the percentage of decrease in price?

1.30 - 1.15 / 130 x 100 = 11.54% decrease

Linear equations involving one variable are solved by calculating the value for the variable in one equation and then substituting this value into the other equation.

Given the equation of $2x + 4 = 6x - 12$. What does x equal?

Move the 2x from one side of the equation to the other, also move the 12 from one side to the other. Remember when moving from one side to the other the quantity being moved becomes a negative of the original amount being moved. This gives us a new equation of: $4 + 12 = 6x - 2x$

Now add each side and get: $16 = 4x$

Now divide by the quantity of x to get your answer: $4 = x$.

Simultaneous equations with 2 unknowns are solved by solving one equation in terms of one of the unknowns and then substituting that value into the other equation.

Example: Solve the following equations

$2x + y = 25$

$x - 2y = -5$

Solve the second one first. Move the -2y to the other side of the equation to make the following: x = -5 + 2y

Now substitute your answer into the other equation to get you answers.

$$2(-5 + 2y) + y = 25$$

$$-10 + 4y + y \quad 25$$

$$5y - 10 = 25$$

$$5y = 35$$

$$y = 7$$

Now go back to the first equation to solve for x using 7 for y.

$$x - 2y = -5$$

$$x - 2(7) = -5$$

$$x - 14 = -5$$

$$x = 9$$

Pythagorean theorem is a basic way of finding what a unknown side of a right triangle is by using the other two sides. The angle opposite the right angle is always known a *c.*

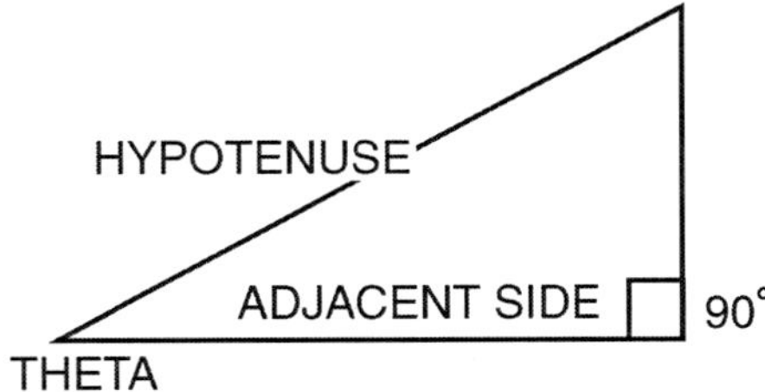

Pythagoreans theory is: $a^2 + b^2 = c^2$

Now lets take a triangle with the two adjoining sides to the right angle being 3 and 4, we need to know what the other side is.

$$3^2 + 4^2 = c^2$$

$$9 + 16 = c^2$$

$$25 = c^2$$

$$5 = c$$

Trigonometric functions are calculated using a simple set of circle formulas similar to the Ohm's Law circle. Using these formulas to get a base answer you plug the base answer into the trigonometric ratios table and you will get the answer to your problem. These three trig circles are:

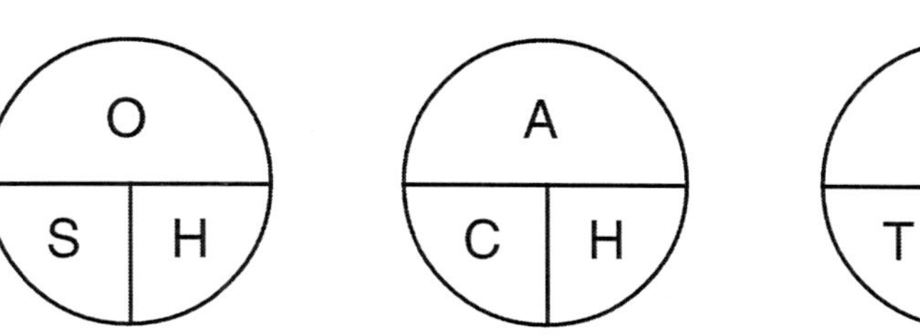

The Trigonometric Radius Chart is found on the Level II Tab.

Lets take a simple problem of a triangle with the following properties. Side a is 20 feet, the angle opposite side b is 40°. What is the hypotenuse?

Because we have the length of side a and the angle opposite of side be we can use the cosine circle to find the answer.

$$cos\ 40 = 20/hyp$$

$$.766\ =\ 20/hyp$$

$$hyp = 20/.766$$

$$hyp\ =\ 26.1\ feet$$

PRACTICAL PROBLEMS

Decimals: 1.   6.774 x .001 =

                   2.   4.986 x .0462 =

                   3.   376.2 x .0022 =

Linear Equations: 6x + 3 - 4x = 13 - 3x

                         4x - 1 =

Simultaneous Equations: 3x + 2y = 16

                               x + 4y = 12

Percentages: 1.   14 is what % of 50

                2.   17 is what % of 29

                3.   26 is what % of 16

Percent Increase or Decrease:

     1.   Halon 1301 increases in price from $6.00/lb to $9.00/lb. What is the percent increase in price?

2. Halon 1301 decreases in price from $9.00/lb to $6.00/lb.

What is the percent decrease in price?

3. Pipe increases in price from $0.84/ft to $0.96/ft.

What is the percent increase?

4. Pipe decreases in price from $11.67/10 feet to $0.92/ft.

What is the percent decrease in price?

Areas & Volumes:

All the questions in this subject are based on a rectangle with sides of 10 and 200 feet.

1. What is the area of the rectangle?

2. The area of the bottm of a tank is 2,000 feet

What is the volume of the tank if it is 12 feet deep?

3. How much air will be left in the tank if we add 32" of water to the tank?

4. How much air is left in the tank if we add another 18" of water?

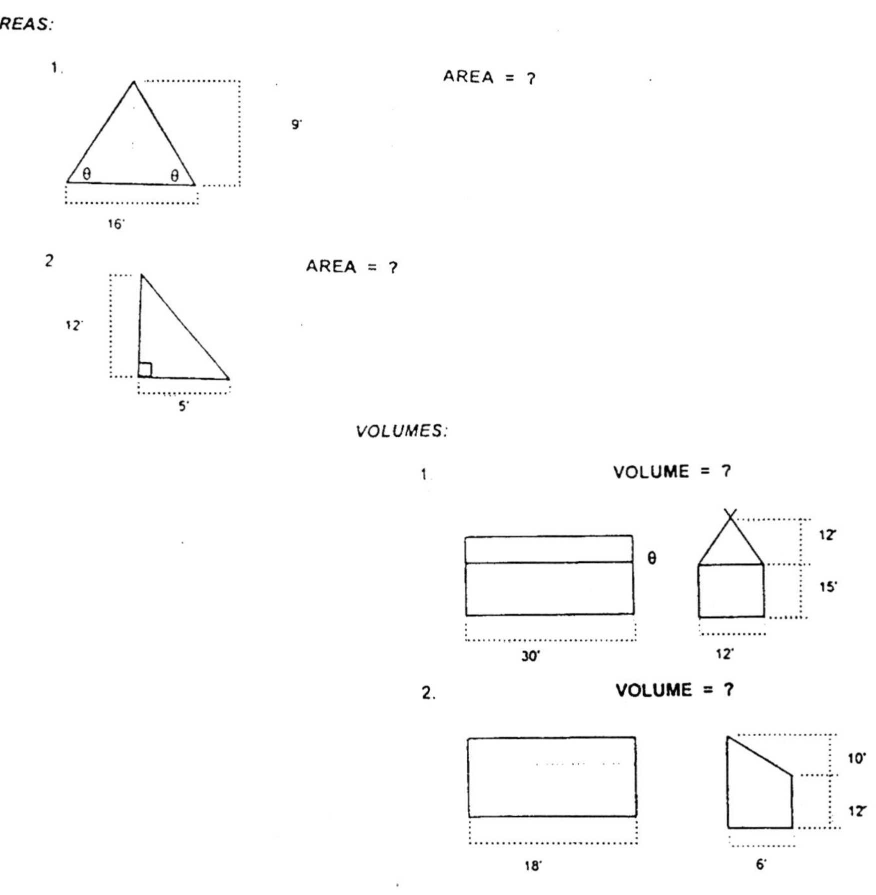

AREA AND VOLUME *(CIRCLES AND CYLINDERS)* :

1. AREA OF CIRCLE = ?

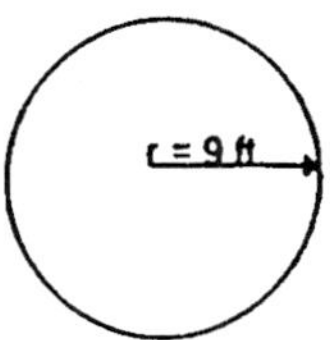

2. AREA OF X-SECTION = ?

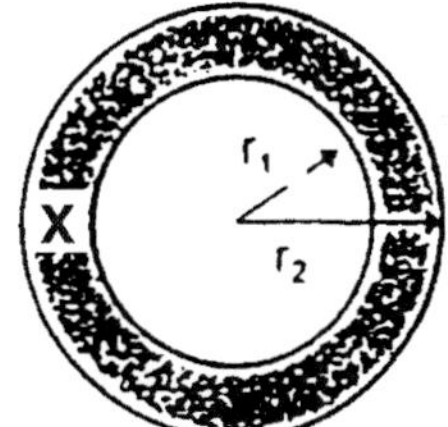

r_1 = 6 in.
r_2 = 8 in.

3. What is the weight of a pipe which is 32 ft. long and has the same cross sectional area as in the previous problem (i.e. 88 In.^{2}). (Steel Weighs 480 lb./ft.^{3})

4. Determine the weight of 50 ft. of pipe with an outside radius of 8" (r_1) and a wall thickness of 1/2". (Steel weighs 480 lb./ft.^{3})

TRIGONOMETRY:

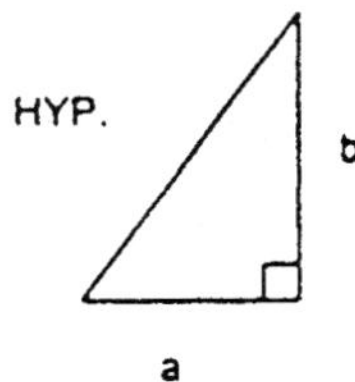

PYTHAGOREAN THEOREM:

$$a^2 + b^2 = \text{Hyp.}^2$$

EXAMPLE: a = 6 ft.
                   b = 12 ft.
                   Hyp. = ?

$$a^2 + b^2 = \text{Hyp.}^2$$
$$(6 \text{ ft.})^2 + (12 \text{ ft.})^2 = (\text{Hyp.})^2$$
$$36 \text{ ft.}^2 + 144 \text{ ft.}^2 = (\text{Hyp.})^2$$
$$180 \text{ ft.}^2 = (\text{Hyp.})^2$$
$$\text{Hyp.} = 13.4 \text{ ft.}$$

PROBLEM: a = 16 ft.
                  Hyp. = 26 ft.
                     b = ?

ANSWERS

Decimals:	1.	0.006774	
	2.	230.4	
	3.	0.8276	
Linear Equation:	1.	7	
Simutaneous Equasions:		1.	$x = 4$
		2.	$y = 2$
Percentages:	1.	28%	
	2.	59%	
	3.	162.5%	
Areas & Volumes:	1.	$2{,}000 \text{ ft}^2$	
	2.	$24{,}000 \text{ ft}^3$	
	3.	$18{,}660 \text{ ft}^3$	
	4.	$15{,}660 \text{ ft}^3$	
Areas:	1.	72 ft^2	
	2.	30 ft^2	
Volumes:	1.	7560 ft^3	
	2.	1836 ft^3	

Area & Volume (circles & cylinders):

1.	254.5 ft^2
2.	88 in^2
3.	9387 lb
4.	4056 lb

Trigonometry: $b = 20.5$ ft

Trigonometric Functions:

1.	Hyp $= 26.2$ ft
2.	$a = 21.45$ ft
3.	$q = 54°$
4.	$q = 40.3°$
5.	$H = 106$ ft

31009 INSTALLATION PRACTICES

REFERENCES:

NFPA 72

DESCRIPTION:

Understand installation methods, proper connections to maintain supervision, proper location of initiating devices, indicating appliances, control panels, annunciators and other system components to comply with fire codes and assure proper operation of the system.

NFPA 72-1999 1-5.5.2.1

All fire alarm systems shall be installed in accordance with the specifications and standards approved by the authority having jurisdiction.

NFPA 72-1999 1-5.5.2.2

Devices and appliances shall be located and mounted such that accidental operation or failure is not caused by vibration or jarring.

NFPA 72-1999 1-5.5.2.3

All fire alarm apparatus requiring rewinding or resetting to maintain normal operation shall be restored to normal as promptly as possible after each alarm and kept in normal condition for operation.

NFPA 72-1999 1-5.5.3

To reduce the possibility of damage caused by induced transients, circuits and equipment shall be protected in accordance with the requirements of NFPA 70, *National Electrical Code*, Article 800.

NFPA 72-1999 1-5.5.4

The installation of wiring, cable, and equipment shall be in accordance with NFPA 70, *National Electrical Code,* and specifically with Article 760, Article 770, and Article 800. Optical fiber cables shall be protected against damage in accordance with Article 760.

NFPA 72-1999 1-5.5.5

All fire alarm systems shall test free of grounds, except parts of circuits or equipment that are intentionally and permanently grounded to provide ground-fault detection, noise suppression, emergency ground signaling, and circuit protection grounding.

NFPA 72-1999 1-5.5.6.1 — Initiating devices shall be selected and installed so as to minimize nuisance alarms.

NFPA 72-1999 1-5.6 — Automatic smoke detection shall be provided at each control unit(s) location to provide notification of fire at that location in areas that are not continuously occupied. Except where ambient conditions prohibit installation of automatic smoke detection, automatic heat detection shall be permitted.

NFPA 72-1999 1-5.7.1.1 — The primary purpose of fire alarm system annunciation is to enable responding personnel to quickly and accurately identify the location of a fire and to indicate the status of emergency equipment or fire safety functions that might affect the safety of occupants. Required annunciation means shall be readily accessible to responding personnel and shall be located as required by the authority having jurisdiction to facilitate an efficient response to the fire situation.

NFPA 72-1999 1-5.7.1.2 — Fire alarm systems serving two or more zones shall identify the zone of origin by annunciation or coded signal.

NFPA 72-1999 1-5.7.2 — Alarm annunciation at the fire command center shall audible and visible.

NFPA 72-1999 1-5.7.3 — For alarm annunciation purposes, each floor of a building shall be considered a separate zone. Where a floor is subdivided by fire or smoke barriers and the fire plan for the protected premises allows relocation of occupants from the zone of origin to another zone on the same floor, each zone on the floor shall be annunciated separately.

NFPA 72-1999 1-5.7.4 — For systems serving more than one building, each building shall be annunciated separately.

NFPA 72-1999 1-5.8.1 — All means of interconnecting equipment, devices, appliances and wiring connections shall be monitored for integrity such that the occurrence of a single open or a single ground-fault condition in the installation conductors or other signaling channels and their restoration to normal shall be automatically indicated within 200 seconds.

NOTE: The provision of a double loop or other multiple path conductor or circuit to avoid electrical monitoring is not acceptable.

NOTE: This code does not have jurisdiction over the monitoring integrity of conductors within equipment, devices, or appliances.

NFPA 72-1999 1-5.8.2

Interconnection means shall be arranged such that a single break or single ground fault does not cause an alarm signal.

NFPA 72-1999 1-5.8.4

An open, ground, or short circuit fault on the installation conductors of one fire alarm notification appliance circuit shall not affect the operation of any other notification circuit.

NFPA 72-1999 1-5.8.5

The occurrence of a wire-to-wire short circuit fault on any fire alarm notification appliance circuit shall result in a trouble signal at the protected premises.

NFPA 72-1999 1-5.8.6.1

Where speakers are used to produce audible fire alarm signals, failure of any audio amplifier or of any tone-generating equipment shall result in an audible trouble signal. Tone-generating and amplifying equipment enclosed as integral parts and serving only a single, listed loudspeaker shall not be required to be monitored.

NFPA 72-1999 1-5.8.6.2

Where a two-way telephone communications circuit is provided, its installation wires shall be monitored for a short circuit fault.

NFPA 72-1999 1-5.8.7.1

Primary and secondary power supplies shall be monitored for the presence of voltage at their point of connection to the fire alarm system. Failure of either supply shall result in a trouble signal in accordance with 1-5.4.6. The trouble signal shall also be visually and audibly indicated at the protected premises. Where the DACT is powered from a protected premises fire alarm system control unit, power failure indication shall be in accordance with this paragraph.

NFPA 72-1999 1-5.8.7.3

The primary power supply failure trouble signal for the DACT shall not be transmitted until the actual battery capacity is depleted by at least 25 percent, but by not more than 50 percent.

NFPA 72-1999 1-6.1.1

The authority having jurisdiction shall be notified prior to installation or alteration of fire alarm equipment or wiring. At the AHJ's request, complete information regarding the system or system alterations, including specifications, wiring diagrams, battery calculations, and floor plans shall be submitted for approval.

NFPA 72-1999 1-6.2.2

Each fire alarm system shall include the following documentation, which shall be delivered to the owner or the owner's representative upon final acceptance of the system.
    (a) An owner's manual and installation instructions covering all system equipment; and
    (b) Record drawings.

NFPA 72-1999 1-6.2.3

For Central Station Fire Alarm System Service, it shall be conspicuously indicated by the prime contractor (see Chapter 4) that the fire alarm system providing service at a protected premises complies with all applicable requirements of this code by providing a means of verification as specified in either 1-6.2.3.1 or 1-6.2.3.2

NFPA 72-1999 1-6.2.3.1

When a Central Station Fire Alarm System is verified by UL®, the installation shall be certificated.

NFPA 72-1999 1-6.2.3.1.1

Central station fire alarm systems providing service that complies with the requirements of this code shall be certificated by the organization that has listed the prime contractor, and a document attesting to this certification shall be located on or near the fire alarm system control unit or, where no control unit exists, on or near a fire alarm system component.

NFPA 72-1999 1-6.2.3.1.2

A central repository of issued certificates, accessible to the authority having jurisdiction, shall be maintained by the organization that has listed the central station.

NFPA 72-1999 1-6.2.3.2

When a Central Station Fire Alarm System is verified by FM®, The installation shall be placarded.

NFPA 72-1999 1-6.2.3.2.1

Central station fire alarm systems providing service that complies with the requirements of this code shall be conspicuously marked by the prime contractor to indicate compliance. The marking shall be by means of one or more securely affixed placards.

NFPA 72-1999 1-6.2.3.2.2

The placard(s) shall be 20 in.^{2} or larger, shall be located on or near the fire alarm system control unit or, where no control unit exists, on or near a fire alarm system component, and shall identify the central station and, where applicable, the prime contractor by name and telephone number.

NFPA 72-1999 1-6.3

A complete and unalterable record of the tests and operations of each system shall be kept until the next test and for 1 year thereafter. The record shall be available for examination and, where required, reported to the authority having jurisdiction. Archiving of records by any means shall be permitted if hard copies of the records can be provided promptly when requested.

NFPA 72-1999 2-1.3.1

Where subject to mechanical damage, an initiating device shall be protected. A mechanical guard used to protect a smoke or heat detector shall be listed for use with the detector being used.

NFPA 72-1999 2-1.3.2

In all cases, initiating devices shall be supported independently of their attachment to the circuit conductors.

NFPA 72-1999 2-1.3.3

Initiating devices shall be installed in all areas where required by the appropriate NFPA standard or the authority having jurisdiction. Each installed initiating device shall be accessible for periodic maintenance and testing.

NFPA 72-1999 2-1.3.4

Duplicate terminals or leads, or their equivalent shall be provided on each initiating device for the express purpose of connecting into the fire alarm system to provide supervision of the connections. Such terminals or leads are necessary to ensure that the wire run is broken and that the individual connections are made to the incoming and outgoing leads or other terminals for signaling and power, except initiating devices that provide equivalent supervision.

NFPA 72-1999 2-1.4.1

Detectors shall not be recessed into the mounting surface in any manner, except where tested and listed for such recessed mounting.

NFPA 72-1999 2-2.2.1

Spot-type heat detectors shall be located on the ceiling not less than 4 in. from the sidewall or on the sidewalls between 4 in. and 12 in. from the ceiling. In the case of solid open joist construction, detectors shall be mounted at the bottom of the joists. In the case of beam construction where beams are less than 12 in. in depth and less than 8 ft on center, detectors shall be permitted to be installed on the bottom of beams.

NFPA 72-1999 2-3.4.3.1 Spot-type smoke detectors shall be located on the ceiling not less than 4 in. from a sidewall to the near edge or, where on a sidewall, between 4 in. and 12 in. down from the ceiling to the top of the detector.

Review the requirements for the location and spacing of spot-type detectors.

Review the requirements for the location and spacing of notification appliances.

31010 BASIC COMMUNICATION SKILLS

REFERENCES:

Basic grammar references.

DESCRIPTION:

Use proper punctuation, vocabulary, spelling, and sentence structure. Follow written instructions.

Use of a comma:

Set off an appositive, that is, an expression that explains or gives additional information about the preceding expression.

Separate the name of a city from the name of a state and enclose the state in commas.

Separate the two main clauses joined by the conjunctions but, and, or, for, neither, nor, or either. The comma precedes the conjunction.

Separate the day of the month from the year.

Separate an introductory word from the rest of the sentence.

When writing figures in thousands, but not in street, room, post office box, zip code, telephone numbers and page numbers.

Set off parenthetical words or phrases.

Do not use a comma to:

Separate two nouns, one of which identifies the other.

Between a name, and of, indicating places or position.

Use a period:

At the end of a declarative or imperative sentence.

After certain initials and abbreviations. The periods are omitted between initials standing for agencies and organizations.

After each letter or number in an outline or itemized list unless the letter or number is enclosed in parentheses.

When an expression in parentheses comes at the end of a sentence and is part of the sentence, place the period outside the parentheses. If the parenthetical expression is independent of the sentence and a period is necessary, place the period within the closing parentheses.

Always place the period inside quotation marks.

Use a semicolon:

May be used to separate the parts of a compound sentence when the comma and conjunction are omitted.

Use to separate long, involved clauses.

In enumeration's of a series of items, use semicolons to separate items that contain commas.

Use before an adverb that serves the purpose of a conjunction

Place the semicolon outside quotation marks

Use it after a closing parenthesis, if the construction of the sentence requires a semicolon. Never use a comma or semicolon before a parenthesis

Use of a colon:

To show a pause between two closely related sentences.

To indicate clock time and to show ratios

In business letter salutations are always followed by a colon

To separate the verse and chapter in biblical references

After a word, phrase, or sentence that introduces lists, a series, tabulations, extracts, tests, and explanations that are in opposition to the introductory words

Do no use a dash with a colon

Use of question marks:

Place a question mark after a direct question but not after an indirect question.

Do not place a question mark after a question that is a request to which no answer is expected.

Use of quotation marks:

Use to enclose the exact words of a speaker or writer. When quoted material is more than one paragraph long place quotation marks at the beginning of each paragraph, but only at the close of the last paragraph.

To enclose the titles of articles, chapter or part titles in books, short poems, songs, papers, and TV and radio shows.

Use of apostrophes:

Use it to indicate the possessive case of nouns.

Do not use it to indicate the possessive case of personal pronouns.

To denote a contraction or omission of letters

To form the plurals of letters and symbols

31011 BASIC METRIC UNITS AND CONVERSIONS

REFERENCES:

ASTM E-380

DESCRIPTION:

Perform conversions to and from basic metric (SI) units.

Be able to perform conversions to and from metric units for the following:

CONVERSION CHARTS

TO CONVERT	INTO	MULTIPLY BY
meters	centimeters	100
meters	millimeters	1,000
meters	kilometers	0.001
meters	inches	39.37
meters	feet	3.261
meters	yards	1.094
yards	meters	.9114
miles(statute)	feet	5,280
miles(statute)	meters	1,609
miles per hour	feet per minute	88
miles per hour	miles per minute	0.1667
pounds per foot	kilos per meter	1.488
square centimeters	square inches	0.1550
square feet	square meters	0.09290
square inches	circular mils	1.273×10^6
square kilometers	square miles	.3861
temperature(°F)-32	temperature (°C)	5/9
watts	BTU per hour	3.413

TIMES NOTES

UNIT	PREFIX	NOTED BY	MULTIPLIER
billion	giga	G	10^9
million	mega	M	10^6
thousand	kilo	K	10^3
thousandth	milli	m	10^{-3}
millionth	micro		10^{-6}
billionth	nano	n	10^{-9}
trillionth	pico	p	10^{-12}
quadrillionth	femto	f	10^{-15}

AREA

1 Square foot	144 square inches	929.03 square centimeters
1 square foot		0.92 square meters
1 square inch		6.4516 square centimeters
1 square yard	9 square feet	0.82 square meters

1 square centimeter	0.155 square inches
1 square meter	10.8 square feet

LENGTH

1 inch	25.4 millimeters
1 inch	2.54 centimeters
1 foot	0.3048 meter
1 yard	0.914 meter
1 mile	1.609 Kilometers

1 millimeter	0.3937 inch
1 centimeter	0.03937 inch
1 meter	3.2808 feet
1 Kilometer	0.6214 miles

WEIGHTS

1 Ounce	28 Grams
1 Gram	0.035 ounce
1 pound	0.45 Kilogram
1 Kilogram	2.2 pounds
1 ton	0.907 metric ton
1 metric ton	1.102 ton

ENERGY

1 JOULE	107 ERGS
1 ft-lb	1.36 Joules
1 BTU	778 ft - lb
1 Calorie	4.18 Joule
1 BTU	252 Calories
1 KW -HR	3415 BTU

POWER

1 H.P. =	550ft-lb/sec	=	33,000ft-lb/min
1 H.P. =	746 watts		
1 KW =	3413BTU/HR		

UNIT SYSTEMS

UNITS	BRITISH	M.K.S.	C.G.S.
Length	Foot (ft)	Meter (M)	Centimeter (cm)
Mass	Slug	Kilogram (KG)	Gram (GM)
Time	Second (sec)	Second (sec)	Second (sec)
Force	Pound (lb)	Newton (nwt)	Dyne
Energy	Foot per pound	Joule per	ERG
		(newton-meter)	(Dyne per cm)
Power	foot pounds/sec	Watt (joule/sec)	ERG/sec
Gravitational Acceleration	32.2ft/sec^2	9.8M/sec^2	980cm/sec^2

Be sure to bring a calculator on exam day (preferably one with metric conversion capability).

32001 PLANS, SPECIFICATIONS, AND CONTRACTS

REFERENCES:

General construction texts and references.

DESCRIPTION:

Be familiar with architectural, mechanical, electrical, structural, and site plans. Review general and special architectural and mechanical specifications. Understand existence of and relationship among contracting parties in the construction industry. Use standard plans and specifications of jobs to determine dimensions, types of materials, elevations, locations and other information needed for a building fire alarm system. Calculate required information from dimensions and other data on the plans and specifications.

Types of drawings used to describe a fire alarm system include:

1) Equipment location drawing

2) Conduit and/or wiring diagram

3) Riser diagram

Specifications define the qualitative requirements for products, materials, and workmanship upon which the contract is based. The specifications are organized in 16 division:

1 - General Requirements

2 - Site Work

3 - Concrete

4 - Masonry

5 - Metals

6 - Wood and Plastic

7 - Thermal and Moisture Protection

8 - Doors and Windows

9 - Finishes

10 - Specialties

11 - Equipment

12 - Furnishings

13 - Special Construction

14 - Conveying Systems

15 - Mechanical

16 - Electrical

17 - Special Systems

Specifications are legal contractual documents. Specifications are generally divided into two parts:

1) General clauses and agreements

2) Technical instructions

General clauses and agreements may contain shipping, handling, and storage of materials, work plans, work rules, and payment schedule.

Technical instructions contain the scope of the project, technical requirements, and design documents. Acceptance of the system details the acceptance test requirements of the system and subsequent certification of the system.

Definitions found in the specifications include:

Contract. The entire agreement between the parties including the form of contract together with the specifications, advertisement, information for bidders, the contractor's proposal (copies of which are bound with or accompany other data), and the contract drawings.

Contract Drawings. Those listed in specifications under the title "contract drawings" and any future changes and revisions.

Contractor. The individual, corporation, company, firm, or other organization which has contracted to furnish the materials and to perform the work under the contract.

Engineer. The chief engineer, acting either personally or through his or her duly authorized representative, who operates within the scope of the authority vested in him or her.

Inspector. Any representative of the engineer designated by the latter to act as inspector within the scope of the authority by the engineer as delegated. The engineer, however, may review and change any decision of an inspector.

Owner. The corporation.

Subcontractor. Anyone other than the contractor performing work at the construction site directly or indirectly for or in behalf of the contractor. Subcontractor does not include someone who furnishes personal services only.

Work. All labor, plant, materials, facilities, and other things necessary or proper for or incidental to the construction of the job.

The principal documents prepared for the installation of fire alarm signaling systems should include drawing and specifications, which contain standard clauses and technical instruction. These documents are the construction contract; which governs the rights, duties, and liabilities of the contractor, the end user, and the engineer who designs, inspects, and tests the work.

Engineering and architectural drawings are called design drawings when they are prepared for bidding purposes. Once the contract is awarded, design drawings become known as contract drawings. Engineering drawings are often supplemented with detail drawings, shop drawings, construction drawings, or working drawings, which are prepared by or for the contractor.

Notes

32003 BASIC PHYSICAL SCIENCE

REFERENCES:

Solutions may involve simple formulas found in basic physics textbooks, but will not involve algebraic manipulations or trigonometry.

DESCRIPTION:

Apply terms, definitions, and concepts from mechanics, electricity, heat, and chemistry.

Fundamentals of chemistry and physics.

What is matter?   Atom vs. molecule.

What are the states of matter?   Solid, liquid, and gas.

What is friction?   Laminar vs. turbulent flow

Types of energy include light, sound, pressure, heat, nuclear, chemical, and electrical.

What represents is the relationship between heat BTU, calorie, heat of fusion, heat of vaporization, and specific and energy?
Heat.

What are the principles of heat and conduction, convection, and radiation energy transfer?

What are the properties of materials?
Mass, density, melting point, triple point, coefficient of expansion, flash point, fire point, auto ignition temperature, flammability, water solubility.

Heat is a form of energy and results from the movement of atoms in a solid, liquid, or gas. Heat can be transferred in three ways:
      (a) Conduction
      (b) Convection
      (c) Radiation

A calorimeter is an instrument for measuring heat or such thermal constants as the specific heat of a substance.

A molecule is the basic and smallest unit of a chemical compound.

Matter exists as a solid, liquid, or gas. A change from one state (allotrope) to another can be brought about by a change in temperature or pressure. The latent heat of vaporization is the energy necessary to separate molecules bound by the attractive forces, which hold them together in the liquid state. There is also a latent heat of fusion required to melt a solid. A gas existing above its critical temperature is known as a permanent gas. A gas existing below its critical temperature is known as a vapor.

Nuclear fission is the breakup of the nucleus of a heavy atom into lighter atoms. The breakup is accompanied by the release of the binding energy. Uncontrolled fission is the source of the colossal energy released in an atom bomb. Fusion is the combining of nuclei to form more massive nuclei.

Combustion fire is rapid persistent chemical changes, which releases heat and light and is accompanied by a flame. Combustion involves the physical processes of the transfer of matter and energy. Combustion is an exothermic, self-sustaining reaction.

If ignition is caused by the introduction of some small external flame, spark, or glowing object, it is considered piloted ignition. If ignition occurs without the assistance of an external pilot source, it is considered auto-ignition.

The ignition temperature of a substance is the minimum temperature at which it will ignite.

32004 FIRE WARNING EQUIPMENT FOR DWELLING UNITS

REFERENCES:

NFPA 72

DESCRIPTION:

Have basic understanding of fire warning equipment for dwellin units, its required protection, power supplies, performance, spacing and location of detectors and alarm sounding devices.

NFPA 72-1999 1-4

Smoke Alarm. A single or multiple station alarm responsive to smoke.

NFPA 72-1999 1-4

Household Fire Alarm System. A system of devices that produces an alarm signal in the household for the purpose of notifying the occupants of the presence of a fire such that they may evacuate the premises.

NFPA 72-1999 1-4

Dwelling Unit. One or more rooms for the permanent use of one or more persons as a space for eating, living, and sleeping, with permanent provisions for cooking and sanitation. For the purposes of this code, *dwelling unit* includes one- and two-family attached and detached dwellings, apartments, and condominiums but does not include hotel and motel rooms and guest suites, dormitories, or sleeping rooms in nursing homes.

NFPA 72-1999 8-1.4.1.6

One- and Two-Family Dwellings. Approved single-station smoke alarms shall be installed in the following locations:

(1) All sleeping rooms (in new construction).

(2) Outside of each separate sleeping area in the immediate vicinity of the sleeping rooms.

(3) On each additional story of the dwelling unit including basements. (Approved smoke alarms powered by batteries shall be permitted in existing construction).

NFPA 72-1999 8-1.4.2

Smoke alarms or detectors shall not be closer than 3 feet from the door to a bathroom or kitchen. Smoke alarms or detectors that are located within 20 feet of a cooking appliance and are equipped with an alarm silencing means or are of the photoelectric type shall be considered acceptable.

NFPA 72-1999 4-3.3

Private Mode Audible Requirements. Audible notification appliances intended for operation in the private mode shall have a sound level of not less than 45dBA at 10ft (3m) or more than 120dBA.

To ensure private mode signals are clearly heard, they shall have a sound level at least 10dBA above average ambient sound level or 5dBA above maximum sound level having a duration of at least 60 seconds, whichever is greater, measured 5 ft (1.5m) above the floor in the occupiable area.

Where audible appliances are installed in sleeping areas, they shall have a sound level of at least 15dBA above average ambient sound level or 5dBA above the maximum sound level having a duration of at least 60 seconds or a sound level of 70dBA, whichever is greater, measured at the pillow level in the occupiable area.

NFPA 72-1999 8-1.4.3

Smoke alarms and smoke detectors shall be capable of detecting abnormal quantities of smoke and shall alarm prior to a gray smoke level of 4 percent per ft. Visible notification appliances located on the ceiling over the bed and within 16 feet of a sleeping occupant shall have a light output rating of 177 cd. Where a visible notification appliance in a sleeping room is mounted more than 24 inches below the ceiling and within 16 feet of the pillow, a minimum rating of 110 cd shall be permitted.

Performance Requirements.

NFPA 72-1999 8-4.1

The two independent power sources shall consist of a primary source that uses the commercial light and power source and a secondary source that consists of a rechargeable battery or standby generator that can operate the system for at least 24 hours in the normal condition followed by 4 minutes of alarm.

NFPA 72-1999 8-4.5

Smoke alarms shall be powered from the commercial light and power source along with a secondary battery source that is capable of operating the device for at least 7 days in normal operation followed by 4 minutes of alarm. Alternatively, smoke alarms shall be powered by a non-replaceable primary battery that is capable of operating the device for at least 10 years followed by 4 minutes of alarm, followed by 7 days of trouble.

NFPA 72-1999 8-4.2

All monitored circuits shall indicate with a distinctive trouble signal the occurrence of a single open or single ground fault.

NFPA 72-1999 8-4.6

Smoke alarms powered by a primary battery capable of operating the device for at least 1 year in normal operation followed by 4 minutes of alarm, followed by at least 7 days of trouble, shall be used only when specifically allowed.

NFPA 72-1999 8-1.4.4

Smoke detectors shall be connected to central controls for power, signal processing, and activation of notification appliances. Smoke alarms shall not be interconnected in numbers that exceed the manufacturer's recommendations, but in no case shall more than 18 initiating devices be interconnected (of which 12 may be smoke alarms) where the interconnecting means is not supervised, nor more than 64 initiating devices be interconnected (of which 42 may be smoke alarms) where the interconnecting means is supervised.

NFPA 72-1999 8-1.2.3

Newly installed fire warning equipment for dwelling units (including self-contained devices) shall produce the audible emergency signal described in ANSI S3.41, *Audible Emergency Evacuation Signal*. (AKA 3-pulse temporal pattern fire alarm evacuation signal)

(a) A fire alarm signal shall take precedence or be clearly recognizable over any other signal even when the non-fire alarm signal is initiated first.

(b) Distinctive alarm signals shall be used so that fire alarms can be distinguished from other functions such as burglar alarms. The use of a common sounding appliance for fire and burglar alarms shall be permitted where distinctive signals are used. *(See 2-2.2.2)*

NFPA 72-1999 8-4.3.1

Means to transmit alarm signals to a constantly attended remote monitoring location shall perform as described in Chapter 5 except that the DACT serving the protected premises only requires a single telephone line and is only required to call a single DACR number. DACT test signals shall be transmitted at least monthly. Such systems shall not be required to be placarded or certificated.

Notes

32005 BASIC INDIVIDUAL SAFETY

REFERENCES:

OSHA Safety and Health Standards

29 CFR Parts 1929/1910

DESCRIPTION:

Follow standard safety practices in performing job tasks. Recognize and call attention to improper safety practices at the work site.

OSHA Publications (202) 219-4667

Seattle Region (206) 553-3212

Portland (503) 378-3272

Notes

32006 FIRST AID PROCEDURES

REFERENCES:

See general handbooks on first aid.

DESCRIPTION:

Understand the basic rules and procedures of first aid.

Remember the ABC's of life support:

A - Airway open

B - Breathing restored

C - Circulation maintained

To control bleeding:

- Apply direct pressure

- Elevate injured limb only if it does not cause further pain or injury

- Apply pressure on supplying artery

- Use a tourniquet only as a last resort to control a life-threatening hemorrhage

Shock is a disturbance in the circulation if blood, which can upset all body functions. It is a dangerous condition and can be fatal. Expect some degree of shock in any emergency. Do not give anything by mouth. Symptoms of shock include unusual weakness or faintness, cold pale clammy skin, rapid weak pulse, shallow irregular breathing, chills, nausea, and unconsciousness.

Treatment for shock:

1) Treat the cause of shock as quickly as possible

2) Maintain an open airway

3) Keep the victim warm and flat. In the case of upper body injury, elevate the head and shoulders 10 inches higher than the feet.

4) Get professional medical help immediately

5) Do not give anything by mouth

Sunstroke (heat stroke) is a suddden onset of high fever caused by sun exposure. Symptoms may include:

Extremely high body temperature (106° F or higher), hot red dry skin, absence of sweating, rapid pulse,

Convulsions, and unconsciousness. Sunstroke is a life-threatening emergency.

Treatment for sunstroke:

1) Get professional medical help immediately

2) Lower body temperature quickly by using cold water

3) Do not give any stimulating beverages such as coffee, tea, or soda

Heat exhaustion (heat prostration) is overexposure to heat or sun. Symptoms may include fatigue, irratability, headaches, faintness, weak rapid pulse, shallow breathing, cold clammy skin, and profuse perspiration.

Treatment for heat exhaustion:

1) Instruct victim to lie down in a cool, shaded area or air-conditioned room and elevate feet.

2) Massage legs toward heart

3) Give cold salt water (1/2 teaspoon salt in 1/2 cup water) or cool sweetened drinks, especially iced tea and coffee every 15 minutes until victim recovers.

4) Do not let victim sit up, even after feeling recovered. Victim should rest for a while longer.

Electrical shock treatment:

1) Remove the victim from electrical contact at once without endangering yourself

2) Determine whether the victim is breathing. If so, keep the victim lying down in a comfortable position protected from cold and monitor carefully.

3) Keep the victim from moving about. After an electrical shock, the heart is very weak. Any sudden effort may result in a heart attack.

4) Do not administer stimulates or opiates. Send for medical help at once.

5) If the victim is not breathing, administer artificial respiration without delay.

NATURAL TRIGONOMETRIC FUNCTIONS

ANGLE	SINE	COSINE	TANGENT	COTANGENT	SECANT	COSECANT	ANGLE
0	0.0000	1.0000	0.0000		1.0000		90
1	0.0175	0.9998	0.0175	56.2900	1.0002	57.2987	89
2	0.0349	0.9994	0.0349	28.6363	1.0006	28.6537	88
3	0.0523	0.9986	0.0524	19.0811	1.0014	19.1073	87
4	0.0698	0.9976	0.0699	14.3007	1.0024	14.3356	86
5	**0.0872**	**0.9962**	**0.0875**	**11.4301**	**1.0038**	**11.4737**	**85**
6	0.1045	0.9945	0.1051	9.5144	1.0055	9.5658	84
7	0.1219	0.9925	0.1228	8.1443	1.0075	8.2055	83
8	0.1392	0.9903	0.1405	7.1154	1.0098	7.1853	82
9	0.1564	0.9877	0.1584	6.3138	1.0125	6.3925	81
10	**0.1736**	**0.9848**	**0.1763**	**5.6713**	**1.0154**	**5.7588**	**80**
11	0.1908	0.9816	0.1944	5.1446	1.0187	5.2408	79
12	0.2079	0.9781	0.2126	4.7046	1.0223	4.8097	78
13	0.2250	0.9744	0.2309	4.3315	1.0263	4.4454	77
14	0.2419	0.9703	0.2493	4.0108	1.0306	4.1336	76
15	**0.2588**	**0.9659**	**0.2679**	**3.7321**	**1.0353**	**3.8637**	**75**
16	0.2756	0.9613	0.2867	3.4874	1.0403	3.6280	74
17	0.2924	0.9563	0.3057	3.2709	1.0457	3.4203	73
18	0.3090	0.9511	0.3249	3.0777	1.0515	3.2361	72
19	0.3256	0.9455	0.3443	2.9042	1.0576	3.0716	71
20	**0.3420**	**0.9397**	**0.3640**	**2.7475**	**1.0642**	**2.9238**	**70**
21	0.3584	0.9336	0.3839	2.6051	1.0711	2.7904	69
22	0.3746	0.9272	0.4040	2.4751	1.0785	2.6695	68
23	0.3907	0.9205	0.4245	2.3559	1.8640	2.5593	67
24	0.4067	0.9135	0.4452	2.2460	1.0945	2.4586	66
25	**0.4226**	**0.9023**	**0.4663**	**2.1445**	**1.1034**	**2.3662**	**65**
26	0.4384	0.8988	0.4877	2.0503	1.1126	2.2812	64
27	0.4540	0.8910	0.5095	1.9626	1.1223	1.2027	63
28	0.4695	0.8829	0.5317	1.8807	1.1326	2.1301	62
29	0.4848	0.8746	0.5543	1.8040	1.1434	2.0627	61
30	**0.5000**	**0.8660**	**0.5774**	**1.7321**	**1.1547**	**2.0000**	**60**
31	0.5150	0.8572	0.6009	1.6643	1.1666	1.9416	59
32	0.5299	0.8480	0.6249	1.6003	1.1792	1.8871	58
33	0.5446	0.8387	0.6494	1.5399	1.1924	1.8361	57
34	0.5592	0.8290	0.6745	1.4826	1.2062	1.7883	56
35	**0.5736**	**0.8192**	**0.7002**	**1.4281**	**1.2208**	**1.7434**	**55**
36	0.5878	0.8090	0.7265	1.3764	1.2361	1.7013	54
37	0.6018	0.7986	0.7536	1.3270	1.2521	1.6616	53
38	0.6157	0.7880	0.7813	1.2799	1.1690	1.6243	52
39	0.6293	0.7771	0.8098	1.2349	1.2868	1.5890	51
40	**0.6428**	**0.7660**	**0.8391**	**1.1918**	**1.3054**	**1.5557**	**50**
41	0.6561	0.7547	0.8693	1.1504	1.3250	1.5243	49
42	0.6691	0.7431	0.9004	1.1106	1.3456	1.4545	48
43	0.6820	0.7314	0.9325	1.0724	1.3673	1.4663	47
44	0.6947	0.7193	0.9657	1.0355	1.3902	1.4396	46
45>	**0.7071>**	**<0.7071**	**1.0000>**	**<1.0000**	**1.4142>**	**<1.4142**	**<45**

Level II Element Review

Element	Title	Reference Documents
33001	Fire Protection Plans and Symbols	NFPA 170 Fire Safety Symbols
33002	Basics of System Layout	NFPA 72 National Fire Alarm Code NFPA 75 Protection of Electronic Computer/Data Processing Equipment
33003	Electrical Installation Standards	NFPA 70 National Electrical Code NFPA 72 National Fire Alarm Code
33004	Basic Fire Alarm Signaling Systems	NFPA 72 National Fire Alarm Code
33005	Supervision & Supervisory Service	NFPA 70 National Electrical Code NFPA 72 National Fire Alarm Code NFPA 101 Life Safety Code Fire Alarm Signaling Systems
33006	Detection Methods	NFPA 72 National Fire Alarm Code NFPA 101 Life Safety Code
33007	Detector Spacing	NFPA 72 National Fire Alarm Code UL Fire Protection Equipment Directory
33008	Power Supplies	NFPA 70 National Electrical Code NFPA 72 National Fire Alarm Code Fire Alarm Signaling Systems
33009	System Acceptance and Periodic Tests	NFPA 72 National Fire Alarm Code Manufacturer's test manuals
33010	Construction Plans	General construction texts and references
33011	Specifications and Cost Estimates	AIA A201
33012	Contracts	General construction texts
33013	Building Codes	NFPA 101 Life Safety Code Standard Building Code Uniform Building Code Fire Protection Through Modern Building Codes Fire Protection Handbook
33014	Insurance Authorities and Their Requirements	
33015	Governmental Agencies	VA Specifications, State, City, and related documents
33016	Protective Premises Fire Alarm Systems	NFPA 72 National Fire Alarm Code NFPA 101 Life Safety Code Fire Alarm Signaling Systems
33017	Auxiliary Fire Alarm Systems	NFPA 72 National Fire Alarm Code Fire Protection Handbook Fire Alarm Signaling Systems

Level II Element Review (continued)

Element	Title	Reference Documents
33018	Supervising Station Fire Alarm Systems	NFPA 72 National Fire Alarm Code NFPA 101 Life Safety Code Fire Alarm Signaling Systems
33019	Proprietary Supervising Station Systems	NFPA 72 National Fire Alarm Code
33020	Central Station Fire Alarm Systems	NFPA 72 National Fire Alarm Code
33021	Manual Fire Alarm Systems and Guard's Tour Service	NFPA 72 National Fire Alarm Code Basic/National Building Code Standard Building Code Uniform Building Code
33022	Heat Sensing Fire Detectors	NFPA 72 National Fire Alarm Code Fire Alarm Signaling Systems Fire Protection Equipment Directory
33023	Smoke Sensing Fire Detectors	NFPA 72 National Fire Alarm Code NFPA 80 Fire Doors and Windows NFPA 90A Air Conditioning and Ventilating Systems Fire Alarm Signaling Systems Fire Protection Equipment Directory Guide for Proper Use of System Smoke Detectors Guide for Proper use of Smoke Detectors in Duct Applications
33024 33025	Radiant Energy Sensing Fire Detectors Sprinkler Waterflow and Supervisory Devices	NFPA 72 National Fire Alarm Code NFPA 13 Installation of Sprinkler Systems NFPA 72 National Fire Alarm Code Fire Protection Handbook Fire Alarm Signaling Systems
33026	Alarm Notification Appliances	NFPA 72 National Fire Alarm Code NFPA 101 Life Safety Code Fire Alarm Signaling Systems Manufacturer's text
33027	Basics of Signal Transmission	NFPA 72 National Fire Alarm Code Related appendix
33028	Business Communications	Basic grammar and writing handbooks
33029	Intermediate Mathematics	Basic textbooks of algebra & trigonometry
34001	Emergency Voice/Alarm Communication Systems	NFPA 72 National Fire Alarm Code NFPA 101 Life Safety Code

Level II Element Review (continued)

Element	Title	Reference Documents
34002	Signal Processing	<u>Training Manual on Fire Alarm Systems</u> <u>Fire Protection Handbook</u> <u>Fire Alarm Signaling Systems</u> Manufacturer's data and electrical and electronic textbooks
34003	Surveys for Fire Alarm and Detection Systems	NFPA 72 National Fire Alarm Code NFPA 101 Life Safety Code <u>Training Manual on Fire Alarm Systems</u> <u>Basic/National Building Code</u> <u>Fire Alarm Signaling Systems</u> <u>Standard Building Code</u> <u>Uniform Building Code</u>
34004	Fire Alarm System Maintenance	NFPA 72 National Fire Alarm Code Electrical textbooks
34005	Fire Alarm System Wiring	NFPA 70 National Electrical Code
34006	Emergency Evacuation Signals	NFPA 72 National Fire Alarm Code
34007	Combination Systems	NFPA 72 National Fire Alarm Code <u>Fire Protection Handbook</u> <u>Fire Alarm Signaling Systems</u>

Notes

33001 FIRE PROTECTION PLANS AND SYMBOLS

REFERENCES:

NFPA 170

DESCRIPTION:

Know the standard design symbols common to the fire alarm industry.

PLACARDS FOR USE BY THE FIRE SERVICE

SYMBOL	Automatic Sprinkler Connection - Siamese	Automatic Sprinkler Connection - Single
CHARACTERISTICS	Square Red Field White Symbol	Square Red Field White Symbol
APPLICATION	For ID of location of a fire department automatic sprinkler connection	For ID of location of a fire department automatic sprinkler connection

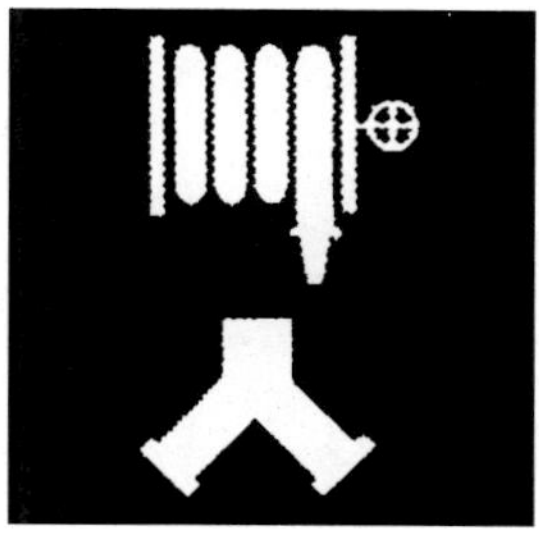

SYMBOL	Fire Department Standpipe Connection	Fire Department Combined Automatic Sprinkler/Standpipe Connection
CHARACTERISTICS	Square Red Field White Symbol	Square Red Field White Symbol
APPLICATION	For ID location of a Fire Department Standpipe connection	For ID location of a Fire Department combined automatic sprinkler/standpipe connection

SYMBOL	Fire Hydrant (all types)	Automatic Sprinkler Control Valve
CHARACTERISTICS	Square Red Field White Symbol	Square Red Field White Symbol
APPLICATION	To ID location of fire hydrant	To ID location of automatic sprinkler control valve

SYMBOL	Electric Panel or Shutoff	Gas Shutoff Valve
CHARACTERISTICS	Square Blue Field White Symbol	Square Red Field White Symbol Green Letter G
APPLICATION	To ID location of electrical panel or other electric shutoff device	To ID location of gas shutoff valve

SYMBOL	Fire Fighting Hose or Standpipe Outlet	Fire Extinguisher
CHARACTERISTICS	Square Blue Field White Symbol	Square Red Field White Symbol
APPLICATION	To ID location of a fire fighting hose or standpipe outlet	To ID the location of a Fire Extinguisher

SYMBOL	Direction Arrow (used with other signs)	Child Care Center
CHARACTERISTICS	Square Red or Blue Field - White Symbol	Square White Field Blue Symbol
APPLICATION	To help direct the way to equipment shown on the accompanying sign	To ID the location of a Child Care Center

<u>ARCHITECTURAL AND ENGINEERING SYMBOLS</u> - See NFPA 170

Notes

33002 BASICS OF SYSTEM LAYOUT

REFERENCES:	DESCRIPTION:
NFPA 72	Know the basic principles of NFPA 72, including requirements for extent of protection, system alarm initiating devices, control functions, alarm indicating appliances, etc.

NFPA 72-1999 1-4

Supplementary. As used in this code, supplementary refers to equipment or operations not required by this code and designated as such by the authority having jurisdiction.

NFPA 72-1999 1-5.1.2

Equipment constructed and installed in accordance with NFPA 72 will be listed for the purpose for which it is used.

NFPA 72-1999 1-5.1.3

NFPA 72-1999A1-5.1.3

Fire alarm system plans and specifications will be developed in accordance with this code by persons who are experienced in the proper design, application, installation, and testing of fire alarm systems. The system designer will be identified on the system design documents. Evidence of qualifications will be provided when requested by the AHJ.

Examples of qualified personnel include individuals who can demonstrate experience on similar systems and have the following qualifications:

- Factory trained and certified on fire alarm systems design.

- NICET certified in fire alarm systems, Level III or higher.

- Licensed or certified by a state or local authority.

NFPA 72-1999 1-5.2.2

All fire alarm system power supplies are to be installed in accordance with NFPA 70, *National Electrical Code*.

NFPA 72-1999 1-5.4.7

Audible alarm notification appliances for fire alarm systems will produce signals that are distinctive from other appliances used for other purposes such as intrusion alarm systems in the same area.

Fire alarm signals will be distinctive in sound from other signals and their sound will not be used for any other purpose.

Supervisory signals will be distinctive in sound from other signals and their sound will not be used for any other purpose. A supervisory signal sound is permitted to be used to indicate a trouble condition. Where the same sound is used for supervisory and trouble signals, the distinction will be by visible annunciation, for example.

Fire alarm, supervisory, and trouble signals will take precedence, in that order over all other signals. Signals from hold-up alarms or other life threatening signals are permitted to take precedence over supervisory and trouble signals, but not fire alarm where acceptable to the authority having jurisdiction.

NFPA 72-1999 1-5.4.8

A means for turning off the alarm notification appliances is permitted where it is key-operated, located within a locked cabinet, or arranged to provide equivalent protection against unauthorized use. Such means will be permitted only if a visible zone alarm indication or the equivalent has been provided and subsequent alarms on other indicating devices or circuits cause the notification appliances to reactivate. A means for turning off alarm notification appliances that is left in the "off" position when there is no alarm will operate an audible trouble signal until the means is restored to normal. Where the AHJ permits automatically turning off the alarm notification appliances, the alarm is not to be turned off in less than 5 minutes. When permitted by the AHJ, the 5-minute requirement will not apply. When permitted by the AHJ, subsequent actuation of another initiating device of the same type in the same space is not required to cause the notification appliance(s) to reactivate.

NFPA 72-1999 1-5.5.2.1

All fire alarm systems are to be installed in accordance with the specifications and standards approved by the authority having jurisdiction.

NFPA 72-1999 1-5.5.2.4

Fire alarm system equipment is to be installed in locations where conditions do not exceed the following upper and lower limits:

- 85% and 110% of nameplate voltage.

- Ambient temperatures of 32°F and 120°F.

- At a relative humidity of 85% at an ambient temperature of 86°F.

Except for equipment specifically listed for use in locations where conditions can exceed these upper and lower limits.

NFPA 72-1999 1-5.6

In areas that are not continuously occupied, automatic smoke detection is to be installed at each fire alarm control unit location to provide notification of fire at that location, except where ambient conditions prohibit the installation of automatic smoke detection, automatic heat detection is permitted.

NFPA 72-1999 2-5.4.3

The location and sensitivity of the detectors will be the result of an engineering evaluation that includes the following:

(1) Structural features, size, and shape of the rooms and bays

(2) Occupancy and uses of the area

(3) Ceiling height

(4) Ceiling shape, surface, and obstructions

(5) Ventilation

(6) Ambient environment

(7) Burning characteristics of the combustible materials present

(8) Configurations of the contents in the area to be protected

NFPA 72-1999 3-8.3.1.2

Except for fire alarm systems dedicated to elevator recall, control, and supervisory service, at least one fire alarm box will be provided to initiate a fire alarm signal for fire alarm systems which include automatic fire detectors or waterflow detection devices, and located where required by the AHJ.

NFPA 72-1999 3-8.3.2.4.2

No more than five waterflow switches are to be connected to a single initiating device circuit.

NFPA 72-1999 3-8.3.3.1.1

No more than twenty supervisory devices are to be connected to a single initiating device circuit.

NFPA 72-1999 3-9.2.1

An auxiliary relay or other listed appliance connected to the fire alarm system used to initiate control of fire safety functions is to be located within 3 feet of the controlled circuit or device. This auxiliary relay or other appliance will operate within the voltage and current limitations of the fire alarm system control unit. The installation wiring between the fire alarm system control unit and the auxiliary relay will be monitored for integrity. Control devices that operate on loss of power or on loss of power to the auxiliary relay will be considered self-monitoring for integrity.

NFPA 72-1999 2-8.2.1

Manual fire alarm boxes are to be distributed throughout a protected area such that they are unobstructed and accessible.

NFPA 72-1999 2-8.2.2

Manual fire alarm boxes are to be installed within 5 feet of an exit doorway opening at each exit on each floor.

NFPA 72-1999 2-8.2.3

Manual fire alarm boxes are to be installed on both sides of group openings over 40 feet in width. Manual fire alarm boxes are to be mounted within 5 feet of either side of the opening.

NFPA 72-1999 2-8.2.4

Additional manual fire alarm boxes are to be installed such that travel distance to the nearest fire alarm box will not be in excess of 200 ft measured horizontally on the same floor.

NFPA 72-1999 1-5.5.2.2

Fire alarm system devices are to be installed such that vibration or jarring does not cause accidental operation or failure.

NFPA 70-1999 110-14(a)

Terminals. Connection of conductors to terminal parts will ensure a thoroughly good connection without damaging the conductors and will be made by means of pressure connectors (including set-screw type), solder lugs, or splices to flexible leads. Connection by means of wire-binding screws or studs and nuts that have upturned lugs or the equivalent will be permitted for No. 10 or smaller conductors.

NFPA 70-1999 110-14(b)

Splices. Conductors will be spliced or joined with splicing devices identified for the use or by brazing, welding, or soldering with a fusible metal or alloy. Soldered splices will first be spliced or joined so as to be mechanically and electrically secure without solder and then soldered. All splices and joints and the free ends of conductors will be covered with an insulation equivalent to that of the conductors or with an insulating device identified for the purpose.

Wire connectors or splicing means installed on conductors for direct burial will be listed for such use.

NFPA 70-1999 300-4

Where subject to physical damage, conductors will be adequately protected.

NFPA 70-1999 760-25

Installation of nonpower-limited fire alarm circuits will be in accordance with NFPA 70-1999 110-3(b), 300-11(a), 300-15, 300-17, and other appropriate articles of Chapter 3, except as provided in Sections 760-26 through 760-30 and except where other articles of NFPA 70-1999 require other methods.

NFPA 70-1999 760-52

Wiring Methods and Materials on Load Side of the power limited fire alarm power source. Fire alarm circuits on the load side of the power source will be permitted to be installed using wiring methods and materials in accordance with either the wiring methods and materials for (a) or (b) as follows:

(a) Non Power Limited Fire Alarm Wiring Methods and Materials. Installation will be in accordance with Section 760-25, and conductors will be solid or stranded copper, except;

The derating factors given in NFPA 70-1999 310-15(b)(2)(a) will not apply, conductors and multiconductor cables described in and installed in accordance with NFPA 70-1999 760-27 and 760-30 will be permitted, and power-limited circuits will be permitted to be reclassified and installed as nonpower-limited circuits if the power-limited fire alarm circuit markings required by NFPA 70-1999 760-42 are eliminated and the entire circuit is installed using the wiring methods and materials in accordance with Part B, Nonpower-Limited Fire Alarm Circuits.

NOTE: Power-limited circuits reclassified and installed as nonpower-limited circuits are no longer power-limited circuits, regardless of the continued connection to a power-limited source.

(b) Power Limited Fire Alarm Wiring Methods and Materials. Power-limited fire alarm conductors and cables described in NFPA 70-1999 760-71 will be installed as detailed in (1), (2), or (3) as follows: Devices will be installed in accordance with NFPA 70-1999 110-3(b), 300-11(a), and 300-15.

1. In raceway or exposed on the surface of ceiling and sidewalls or fished in concealed spaces. Cable splices or terminations will be made in listed fittings, boxes, enclosures, fire alarm devices, or utilization equipment. Where installed exposed cables will be adequately supported and installed in such a way that maximum protection against physical damage is afforded by building construction such as baseboards, door frames, ledges, etc. Where located within 7 feet of the floor, cables will be securely fastened in an approved manner at intervals of not more than 18 inches.

2. In metal raceways or rigid nonmetallic conduit where passing through a floor or wall to a height of 7 feet above the floor, unless adequate protection can be afforded by building construction such as detailed in (1), or unless an equivalent solid guard is provided.

3. In rigid metal conduit, rigid nonmetallic conduit, intermediate metal conduit, or electrical metallic tubing where installed in hoistways; except as provided for in NFPA 70-1999 620-21 for elevators and similar equipment; and other wiring methods and materials installed in accordance with the requirements of NFPA 70-1999 760-3 will be permitted to extend or replace the conductors and cables described in NFPA 70-1999 760-71 and permitted by NFPA 70-1999 760-52(b).

NFPA 72-1999 2-1.3.1

Where subject to mechanical damage, an initiating device will be protected by a mechanical guard listed for use with the detector being used.

NFPA 72-1999 4-4.4.2.1

NFPA 72-1999 Table 4-4.4.2.1 Corridor Spacing for Visible Appliances, applies to corridors not exceeding 20 feet in width. For corridors greater than 20 feet wide, refer to NFPA 72-1999 Figure 4-4.4.1.1 and NFPA 72-1999 Tables 4-4.4.1.1(a) and (b). In a corridor application, visible appliances will be rated at not less than 15 candela.

NFPA 72-1999 4-4.4.2.2

Visible notification appliances will be located no more than 15 ft from the end of a corridor with a separation no greater than 100 ft between visible notification appliances. Where there is an interruption of the concentrated viewing path, such as a fire door, an elevation change, or any other obstruction, the area will be considered as a separate corridor.

Review the operational characteristics of the different types of automatic detectors. (Refer to the appropriate sections of this handbook.)

33003 ELECTRICAL INSTALLATION STANDARDS

REFERENCES:

NFPA 70

NFPA 72

DESCRIPTION:
Know the National Electrical Code as it applies to installation of fire alarm systems, including automatic fire detection systems, and fire suppression systems. Understand power-limited and nonpower-limited circuits and wiring.

NFPA 70-1999 760-1

This article covers the installation of fire alarm system wiring and equipment including all circuits controlled and powered by the fire alarm system.

NFPA 70-1999 760-2

Fire Alarm Circuit. The portion of the wiring system between the load side of the overcurrent device or power-limited supply and the connected equipment of all circuits powered and controlled by the fire alarm system. Fire alarm circuits are classified as either nonpower-limited or power-limited.

Fire Alarm Circuit Integrity (CI) Cable. Cable used in fire alarm systems to ensure continued operation or the survivability of critical circuits during a specified time under fire conditions.

NFPA 70-1999 760-10

Fire alarm circuits are to be identified at terminal and junction locations in a manner, which will prevent interference with the signaling circuit during testing and servicing.

NFPA 70-1999 760-23

Nonpower-Limited Fire Alarm Circuit Overcurrent Protection. Overcurrent protection for conductors No. 14 and larger will be provided in accordance with the conductor ampacity without applying the derating factors of NFPA 70-1999 310-15 to the ampacity calculation. Overcurrent protection will not exceed 7 amperes for No. 18 conductors and 10 amperes for No. 16 conductors.

NFPA 70-1999 760-24

Nonpower-Limited Fire Alarm Circuit Overcurrent Device Location. Overcurrent devices will be located at the point where the conductor to be protected receives its supply, with some exceptions.

NFPA 70-1999 760-26

Conductors of Different Circuits in Same Cable, Enclosure, or Raceway.

Class 1 with NPLFA Circuits. Class 1 and nonpower-limited fire alarm circuits are permitted to occupy the same cable, enclosure, or raceway without regard to whether the individual circuits are alternating current or direct current, provided all conductors are insulated for the maximum voltage of any conductor in the enclosure or raceway.

Fire Alarm with Power-Supply Circuits. Power-supply and fire alarm circuit conductors will be permitted in the same cable, enclosure, or raceway only where connected to the same equipment.

NFPA 70-1999 760-27

NPLFA Circuit Conductors

Sizes and Use. Only copper conductors will be permitted to be used for fire alarm systems. No. 18 conductors and No. 16 conductors will be permitted to be used, provided they supply loads that do not exceed the ampacities given in NFPA 70-1999 Table 402-5 and are installed in a raceway, an approved enclosure, or a listed cable. Conductors larger than No. 16 will not supply loads greater than the ampacities given in NFPA 70-1999 310-15, as applicable.

Insulation. Insulation on conductors will be suitable for 600 volts. Conductors larger than No. 16 will comply with NFPA 70-1999 310. Conductors in sizes No. 18 and 16 will be Type KF-2, KFF-2, PAFF, PTFF, PF, PFF, PGF, PGFF, RFH-2, RFHH-2, RFHH-3, SF-2, SFF-2, TF, TFF, TFN, TFFN, ZF, or ZFF. Conductors with other types and thickness of insulation will be permitted if listed for nonpower-limited fire alarm circuit use. Wire Types PAF and PTF will be permitted only for high-temperature applications between 90°C (194°F) and 250°C (482°F).

NOTE: For application provisions, see NFPA 70-1999 Table 402-3.

Conductor Materials. Conductors will be solid or stranded copper.

NFPA 70-1999 760-30

Multiconductor NPLFA Cables. Multiconductor nonpower-limited fire alarm cables that meet the requirements of NFPA 70-1999 760-31 is permitted to be used on fire alarm circuits operating at 150 volts or less and will be installed in accordance with (a) and (b).

(a) NPLFA Wiring Method. Multiconductor nonpower-limited fire alarm circuit cables will be installed as follows.

1. In raceway or exposed on surface of ceiling and sidewalls or fished in concealed spaces. Cable splices or terminations will be made in listed fittings, boxes, enclosures, fire alarm devices, or utilization equipment. Where installed, exposed cables will be adequately supported and installed in such a way that maximum protection against physical damage is afforded by building construction such as baseboards, door frames, ledges, etc. Where located within 7 feet of the floor, cables will be securely fastened in an approved manner at intervals of not more than 18 inches.

2. In metal raceway or rigid nonmetallic conduit where passing through a floor or wall to a height of 7 feet above the floor unless adequate protection can be afforded by building construction such as detailed in (1), or unless an equivalent solid guard is provided.

3. In rigid metal conduit, rigid nonmetallic conduit, intermediate metal conduit, or electrical metallic tubing where installed in hoistways, except as provided for in NFPA 70-1999 620-21 for elevators and similar equipment.

(b) Applications of Listed NPLFA Cables. The use of nonpower-limited fire alarm circuit cables will comply with (1) through (4).

(1) Ducts and Plenums. Multiconductor nonpower-limited fire alarm circuit cables, Types NPLFP, NPLFR, and NPLF, will not be installed exposed in ducts or plenums. See NFPA 70-1999 300-22(b).

(2) Other Spaces Used for Environmental Air. Cables installed in other spaces used for environmental air will be Type NPLFP, except NPLFR and NPLF cables that are installed in compliance with NFPA 70-1999 300-22(c), and other wiring methods in accordance with NFPA 70-1999 300-22(c) and conductors in compliance with NFPA 70-1999 760-27(c).

(3) Riser. Cables installed in vertical runs and penetrating more than one floor or cables installed in vertical runs in a shaft will be Type NPLFR. Floor penetrations requiring Type NPLFR will contain only cables suitable for riser or plenum use, except type NPLF or other cables specified in NFPA 70-1999 Chapter 3 that are in compliance with NFPA 70-1999 760-27(c) and enclosed by metal raceway, and type NPLF cables located in a fireproof shaft having firestops at each floor.

NOTE: See NFPA 70-1999 300-21 for floor penetration firestop requirements.

(4) Other Wiring Within Buildings. Cables installed in building locations other than the locations covered in NFPA 70-1999 760-30(b)(1), (2), and (3) will be Type NPLF, except NFPA 70-1999 Chapter 3 wiring

methods with conductors in compliance with NFPA 70-1999 760-27(c). Type NPLFP or Type NPLFR cables will be permitted.

NFPA 70-1999 760-41

Power Sources for Power Limited Fire Alarm Circuits. The power source for a power-limited fire alarm circuit will be as specified in (a), (b), or (c):

NOTE: NFPA 70-1999 Tables 12(a) and 12(b) in Chapter 9 provide the listing requirements for power-limited fire alarm circuit sources.

(a) Transformers. A listed PLFA or Class 3 transformer.

(b) Power Supplies. A listed PLFA or Class 3 power supply.

(c) Listed Equipment. Listed equipment marked to identify the PLFA power source.

NOTE: Examples of listed equipment are a fire alarm control panel with integral power source; a circuit card listed for use as a PLFA source, where used as part of a listed assembly; a current-limiting impedance, listed for the purpose or part of a listed product, used in conjunction with a nonpower-limited transformer or a stored energy source, e.g., storage battery, to limit the output current.

NFPA 70-1999 760-54

Installation of Conductors and Equipment.

Separation from Electric Light, Power, Class 1, NPLFA, and Medium Power Network-Powered Broadband Communications Circuit Conductors.

(1) In Cables, Compartments, Enclosures, Outlet Boxes, or Raceways. Power-limited circuit cables and conductors will not be placed in any cable, cable tray, compartment, enclosure, outlet box, raceway, or similar fitting with conductors of electric light, power, Class 1, nonpower-limited fire alarm circuit conductors, or medium power network-powered broadband communications circuits, except where the conductors of the electric light, power, Class 1, nonpower-limited fire alarm, and medium power network-powered broadband communications circuit conductors are separated by a barrier from the power-limited fire alarm circuits. In enclosures, power-limited fire alarm circuits are permitted to be installed in a raceway within the enclosure to separate them from Class 1, electric light, power, nonpower-limited fire alarm, and medium power network-powered broadband communications circuits, *except* conductors in compartments, enclosures, device boxes, outlet boxes, or similar fittings, where electric light, power, Class 1, nonpower-limited fire alarm, or medium power network-powered broadband communications circuit conductors are introduced solely to connect to the equipment connected to power-limited circuits to which the other conductors are connected, and;

a. The electric light, power, Class 1, nonpower-limited fire alarm, and medium power network-powered broadband communications circuit conductors are routed to maintain a minimum of 0.25 inch separation from the conductors and cables of power-limited fire alarm circuits, or;

b. The circuit conductors operate at 150 volts or less to ground and also comply with one of the following:

1. The fire alarm power-limited circuits are installed using Types FPL, FPLR, FPLP or permitted substitute cables, provided these power-limited cable conductors extending beyond the jacket are separated by a minimum of 0.25 inch or by a nonconductive sleeve or nonconductive barrier from all other conductors, or

2. The fire alarm power-limited circuit conductors are installed as nonpower-limited fire alarm circuits in accordance with NFPA 70-1999 760-25 *and except* conductors entering compartments, enclosures, device boxes, outlet boxes, or similar fittings, where electric light, power, Class 1, nonpower-limited fire alarm, or medium power network-powered broadband communications circuit conductors are introduced solely to connect to the equipment connected to power-limited fire alarm circuits or to other circuits controlled by the fire alarm system to which the other conductors in the enclosure are connected. If the conductors must enter an enclosure that is provided with a single opening, they will be permitted to enter through a single fitting (such as a tee) provided the conductors are separated from the conductors of the other circuits by a continuous and firmly fixed nonconductor, such as flexible tubing.

(2) In Hoistways. Power-limited fire alarm circuit conductors will be installed in rigid metal conduit, intermediate metal conduit, or electrical metallic tubing in hoistways, except as provided for in NFPA 70-1999 620-21 for elevators and similar equipment.

(3) Other Applications. Power-limited fire alarm circuit conductors will be separated at least 2 in. (50.8 mm) from conductors of any electric light, power, Class 1, nonpower-limited fire alarm, or medium power network-powered broadband communications circuits, *except* where either;

(1) all of the electric light, power, Class 1, nonpower-limited fire alarm, and medium power network-powered broadband communications circuit conductors or;

(2) all of the power-limited fire alarm circuit conductors are in raceway or in metal-sheathed, metal-clad, nonmetallic-sheathed, or Type UF cables.

and except where all of the electric light, power, Class 1, nonpower-limited fire alarm, and medium power network-powered broadband communications circuit conductors are permanently separated from all of the power-limited fire alarm circuit conductors by a continuous and firmly fixed nonconductor, such as porcelain tubes or flexible tubing in addition to the insulation on the wire.

(a) Conductors of Different PLFA Circuits, Class 2, Class 3, and Communications Circuits in Same Cable, Enclosure, or Raceway.

(1) Two or More PLFA Circuits. Cable and conductors of two or more power-limited fire alarm circuits, communications circuits, or Class 3 circuits will be permitted in the same cable, enclosure, or raceway.

(2) Class 2 Circuits with PLFA Circuits. Conductors of one or more Class 2 circuits will be permitted within the same cable, enclosure, or raceway with conductors of power-limited fire alarm circuits, provided that the insulation of the Class 2 circuit conductors in the cable, enclosure, or raceway is at least that required by the power-limited fire alarm circuits.

(3) Low Power Network-Powered Broadband Communications Cables and PLFA Cables. Low power network-powered broadband communications circuits will be permitted in the same enclosure or raceway with PLFA cables.

(b) Support of Conductors. Power-limited fire alarm circuit conductors will not be strapped, taped, or attached by any means to the exterior of any conduit or other raceway as a means of support.

(c) Conductor Size. Conductors of No. 26 will only be permitted where spliced with a connector listed as suitable for No. 26 to No. 24 or larger conductors that are terminated on equipment or where the No. 26 conductors are terminated on equipment listed as suitable for No. 26 conductors. Single conductors will not be smaller than No. 18.

NFPA 70-1999 760-61

Applications of Listed PLFA Cables

Power-limited fire alarm cables will comply with (a) through (c) or, where cable substitution is being made, with (d).

(a) Plenum. Cables installed in ducts, plenums, and other spaces used for environmental air will be Type FPLP, except Types FPLP, FPLR, and FPL cable installed in compliance with NFPA 70-1999 300-22.

(b) Riser. Cables installed in vertical runs and penetrating more than one floor or cables installed in vertical runs in a shaft will be Type FPLR. Floor penetrations requiring Type FPLR will contain only cables suitable for riser or plenum use, except where the cables are installed in metal raceway or are located in a fireproof shaft having firestops at each floor, and except Type FPL in one- and two-family dwellings.

NOTE: See NFPA 70-1999 300-21 for firestop requirements for floor penetrations.

(c) Other Wiring Within Buildings. Cables installed in building locations other than the locations covered in (a) and (b) will be Type FPL, except where the cables are enclosed in raceway, and except cables specified in NFPA 70-1999 Chapter 3 that meet the requirements of NFPA 70-1999 760-71(a) and (b) and are installed in nonconcealed spaces where the exposed length of cable does not exceed 10 feet, and except a portable fire alarm system provided to protect a stage or set when not in use is permitted to use wiring methods in accordance with NFPA 70-1999 530-12.

(d) Cable Uses and Permitted Substitutions. The uses and permitted substitutions for power-limited fire alarm circuit cables listed in NFPA 70-1999 Table 760-61 will be considered suitable for the purpose and is permitted.

NOTE: For information on multipurpose cables (Types MPP, MPR, MPG, MP) and communications cables (Types CMP, CMR, CMG, CM), see NFPA 70-1999 800-50.

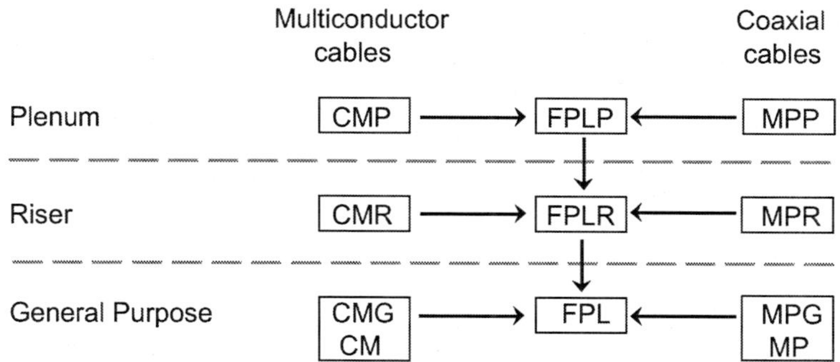

Type CM --Communications wires and cables
Type FPL --Power-limited fire alarm cables
Type MP--Multipurpose cables (coaxial cables only)

A→B Cable A shall be permitted to be used in place of cable B.
No. 26 minimum

Figure 760-61 Cable substitution hierarchy

| Cable Type | Use | References | Permitted Substitutions | |
			Multi-conductor	Coaxial
FPLP	Power limited fire alarm plenum cable	760-61(a)	CMP	MPP
FPLR	Power-limited fire alarm riser cable	760-61(b)	CMP, FPLP, CMR	MPP, MPR
FPL	Power-limited fire alarm cable	760-61(c)	CMP, FPLP, CMR, FPLR, CMG, CM	MPP, MPR, MPG, MP

Table 760-61. Cable Uses and Permitted Substitutions

NFPA 70-1999 760-71

Listing and Marking of PLFA Cables and Insulated Continuous Line-Type Fire Detectors.

Type FPL cables installed as wiring within buildings will be listed as being resistant to the spread of fire and other criteria in accordance with (a) through (h) and will be marked in accordance with (i). Insulated continuous line-type fire detectors will be listed in accordance with (j).

(a) Conductor Materials. Conductors will be solid or stranded copper.

(b) Conductor Size. The size of conductors in a multiconductor cable will not be smaller than No. 26. Single conductors will not be smaller than No. 18.

(c) Ratings. The cable will have a voltage rating of not less than 300 volts.

(d) Type FPLP. Type FPLP power-limited fire alarm plenum cable will be listed as being suitable for use in ducts, plenums, and other space used for environmental air and will also be listed as having adequate fire-resistant and low smoke-producing characteristics.

NOTE: One method of defining low smoke-producing cable is by establishing an acceptable value of the smoke produced when tested in accordance with the Standard Method of Test for Fire and Smoke Characteristics of Wires and Cables, NFPA 262-1994, to a maximum peak optical density of 0.5 and a maximum average optical density of 0.15. Similarly, one method of defining fire-resistant cables is by establishing maximum allowable flame travel distance of 5 feet when tested in accordance with the same test.

(e) Type FPLR. Type FPLR power-limited fire alarm riser cable will be listed as being suitable for use in a vertical run in a shaft or from floor to floor and will also be listed as having fire-resistant characteristics capable of preventing the carrying of fire from floor to floor.

NOTE: One method of defining fire-resistant characteristics capable of preventing the carrying of fire from floor to floor is that the cables pass the requirements of the Standard Test for Flame Propagation Height of Electrical and Optical-Fiber Cable Installed Vertically in Shafts, ANSI/UL 1666-1997.

(f) Type FPL. Type FPL power-limited fire alarm cable will be listed as being suitable for general-purpose fire alarm use, with the exception of risers, ducts, plenums, and other spaces used for environmental air and will also be listed as being resistant to the spread of fire.

NOTE: One method of defining resistant to the spread of fire is that the cables do not spread fire to the top of the tray in the vertical-tray flame test in the Reference Standard for Electrical Wires, Cables and Flexible Cords, ANSI/UL 1581-1991. Another method of defining resistant to the spread of fire is for the damage (char length) not to exceed 4 feet 11 inches. when performing the CSA vertical flame test — cables in cable trays, as described in Test Methods for Electrical Wires and Cables, CSA C22.2 No. 0.3-M-1985.

(g) Fire Alarm Circuit Integrity (CI) Cable. Cables suitable for use in fire alarm systems to ensure survivability of critical circuits during a specified time under fire conditions will be listed as circuit integrity (CI) cable. Cables identified in NFPA 70-1999 760-71(d), (e), and (f) meeting the requirements for circuit integrity will have the additional classification using the suffix "CI" (for example, FPLP-CI, FPLR-CI and FPL-CI).

NOTE No. 1: This cable is used for fire alarm circuits as one method of complying with the survivability requirements of the National Fire Alarm Code, NFPA 72-1999, 3-8.4.1.1, 3-4.2.2.2, 3-8.4.1.3.3, and 3-8.4.1.3.3.3, that the cable maintain its electrical function during fire conditions for a defined period of time.

NOTE No. 2: One method of defining circuit integrity (CI) cable is by establishing a minimum 2-hour fire resistance rating for the cable when tested in accordance with the Standard for Tests of Fire Resistive Cables, UL 2196-1995.

(h) Coaxial Cables. Coaxial cables will be permitted to use 30 percent conductivity copper-covered steel center conductor wire and will be listed as Type FPLP, FPLR, or FPL cable.

(i) Cable Marking. The cable will be marked in accordance with NFPA 70-1999 Table 760-71(i). The voltage rating will not be marked on the cable. Cables that are listed for circuit integrity will be identified with the suffix CI as defined in NFPA 70-1999 760-71(g).

NOTE: Voltage ratings on cables may be misinterpreted to suggest that the cables may be suitable for Class 1, electric light, and power applications. Voltage markings will be permitted where the cable has multiple listings and voltage marking is required for one or more of the listings.

Cable Marking	Type	Listing References
FPLP	Power-limited fire alarm plenum cable	**NFPA 70-1999 760-71(d) and (i)**
FPLR	Power-limited fire alarm riser cable	**NFPA 70-1999 760-71(e) and (i)**
FPL	Power-limited fire alarm cable	**NFPA 70-1999 760-71(f) and (i)**

NFPA 70-1999 Table 760-71(i). Cable Markings

NOTE: Cables identified in (d), (e), and (f) meeting the requirements for circuit integrity will have the additional classification using the suffix "CI" (for example, FPLP-CI, FPLR-CI, and FPL-CI).

NOTE: Cable types are listed in descending order of fire-resistance rating.

(j) Insulated Continuous Line-Type Fire Detectors. Insulated continuous line-type fire detectors will be rated in accordance with (c), listed as being resistant to the spread of fire in accordance with (d) through (f), marked in accordance with (i), and the jacket compound will have a high degree of abrasion resistance.

33004 BASIC FIRE ALARM SIGNALING SYSTEMS

REFERENCES:

NFPA 72

DESCRIPTION:

Know the various types of fire alarm signaling systems. Understand the operation and differences between noncoded, zone noncoded, coded, master coded, march time coded, selective coded, zone coded, voice alarm communication and the recommended evacuation signaling systems.

NEMA 2.3

NONCODED SYSTEM. A noncoded system is one in which a continuous fire alarm signal is transmitted for a predetermined period of time after which the alarm notification appliances may be manually or automatically restored to normal.

NOTE: Restoring the system to normal, following a fire alarm, may involve the resetting of detection devices and the control unit, and is implied in all of the following descriptions.

NEMA 2.4

ZONED NONCODED SYSTEM. A zoned noncoded system provides for the continuous transmission of the fire alarm signal as in a noncoded system, but also incorporates an annunciator or display to indicate the exact area, or zone, of the building from which the alarm originated.

Before continuing with the description of a variety of coded system types, it should be recognized that coded systems use a coding device, which may be electromechanical or electronic in nature, to produce a predetermined, patterned, and distinctive fire alarm signal. The coding mechanism may be integral to the control unit, or it may be employed in a coded manual fire alarm box, or it may be used in a code transmitter device to which are connected noncoded manual and automatic fire detection devices.

NEMA 2.5

CODED SYSTEM. A coded system is one in which not less than three rounds of coded fire alarm signals are transmitted after which the fire alarm system may be manually or automatically restored to normal. (Also see "selective coded system").

NEMA 2.6

MASTER CODED SYSTEM. A master coded system is one in which the coding mechanism provides for the transmission of the same common-coded fire alarm signal in response to the initiation of an alarm from any location in the building. Common code (e.g. 4-4) should be selected to produce a distinctive pulsing of the system's alarm notification appliances so that building occupants know unmistakably that a fire condition exists.

NEMA 2.7

MARCH TIME CODED SYSTEM. A march time coded system is actually a noncoded system that operates much like the master coded system except that the coding mechanism is arranged to produce a fire alarm signal at a march time rate of approximately 120 pulses per minute.

NEMA 2.8

SELECTIVE CODED SYSTEM. A selective coded system is one in which a number of code transmitters or remotely mounted devices, or both, are used; the coding mechanism for each coded unit is systematically selected and organized to produce its own distinctive coded alarm signal identifying its location (floor or zone) in the system. In this way, the sounding of a fire alarm signal notifies the building occupants of an alarm of fire, and, at the same time, tells key building personnel, such as an in-house fire brigade, of the location from which the alarm was initiated. Normally, four rounds of the selective code are sounded.

NEMA 2 9

ZONED CODED SYSTEM. A zoned coded system also alerts both the building occupants and key building personnel. While the audible alarms in a zoned coded system function identically to those in a selective coded system, zoned coded systems generally are less specific about the origin of the signal. As its name implies, the signal denotes only the general area or zone from which one or more different detection devices may have initiated the alarm.

NFPA 72-1999 A-3-8.4.1.2

NFPA 72-1999 1-5.4.7 requires that fire alarm signals be distinctive in sound from other signals and that this sound not be used for any other purpose. To meet this requirement, the fire alarm signal used to notify building occupants of the need to evacuate (leave the building) will be in accordance with ANSI S3.41, Audible Emergency Evacuation Signal.

NFPA 72-1999
3-8.4.1.3.5.3.1

Voice/Alarm Signaling Service Functional Sequence.
In response to an initiating signal indicative of a fire emergency, the system will automatically transmit, either immediately or after a delay acceptable to the authority having jurisdiction, the following:

(a) If the emergency voice/alarm communications service is used to transmit a voice evacuation message, the voice message will be preceded and followed by a minimum of two cycles of the audible emergency evacuation signal specified in NFPA 72-1999 3-8.4.1.2

(b) If the emergency voice/alarm communications service is used to transmit relocation instructions or other non-evacuation messages, a continuous alert tone of 3-second to 10-second duration followed by message (or messages where multichannel capability is provided) will be repeated at least three times to direct the occupants of the alarm signal initiation zone and other zones in accordance with the building's fire evacuation plan.

(c) An evacuation signal will be transmitted to the alarm signal initiation zone and other zones in accordance with the building's fire evacuation plan.

If emergency voice/alarm communications are used to notify all occupants automatically and simultaneously to evacuate a protected premises during a fire emergency, and the functional sequence described in NFPA 72-1999 3-8.4.1.3.5.3.1(a) is provided, the capability to notify portions of the protected premises selectively is not required but, if required, will meet the requirements of NFPA 72-1999 3-8.5.1.3.

NFPA 72-1999
3-8.4.1.3.5.3.3

Live voice instructions will override all previously initiated signals on that channel and will have priority over any subsequent automatically initiated signals on that channel. If multichannel application is required, subsequent alarms will be activated in accordance with NFPA 72-1999 3-8.4.1.3.5.2.

If emergency voice/alarm communications are used to notify all occupants automatically and simultaneously to evacuate the protected premises during a fire emergency, the ability to give live voice instructions is not required but, if provided, will meet the requirements of

Notes

33005 SUPERVISION AND SUPERVISORY SERVICE

REFERENCES:

NFPA 70

NFPA 72

NFPA 101

*Fire Alarm
Signaling Systems*

DESCRIPTION:

Know the terms supervision, monitoring integrity, supervisory service, alarm signal, supervisory signal and trouble signal. Understand the basic supervision requirements of a fire alarm system, its purpose, and the parts of the system requiring supervision, those excluded and why, how supervision is achieved, what supervisory services are available, their purpose, and the type and location of each of the indicating signals listed above.

Supervision. The term supervised refers to monitoring of the circuit, switch, or device in such a manner that a trouble signal is received when a fault that would prevent normal operation of the system occurs.

Supervisory Service. The service required to monitor performance of guard patrols and the operative condition of automatic sprinkler systems and of other systems for the protection of life and property.

Supervisory Signal. A signal indicating the need of action in connection with the supervision of guards' tours, sprinkler and other extinguishing systems or equipment, or with the maintenance features of other protective systems.

Trouble Signal. An audible signal indicating trouble of any nature, such as a circuit break or ground, occurring in the devices or wiring associated with a protective signaling system.

Alarm Signal. A signal indicating an emergency requiring immediate action, as an alarm for fire from a manual box, a water flow alarm, an alarm from an automatic fire alarm system, or other emergency signal.

NFPA 72-1999 3-4.2

Initiating Device Circuits (IDC), Notification Appliance Circuits (NAC), and Signaling Line Circuits (SLC) will be designated by class or style, or both, dependent upon the circuit's ability to operate during specified fault conditions.

NFPA 72-1999 1-5.8.1

All means of interconnecting equipment, devices, appliances, and wiring connections will be monitored for the integrity of the interconnecting conductors or equivalent path such that the occurrence of a single open or a single ground-fault condition in the installation conductors or other signaling channels and their restoration to normal will be automatically indicated within 200 seconds;

except styles of initiating device circuits, signaling line circuits, and notification appliance circuits tabulated in NFPA 72-1999 Tables 3-5, 3-6, and 3-7 that do not have an "X" under "Trouble" for the abnormal condition indicated, and;

except shorts between conductors other than as required by NFPA 72-1999 1-5.8.5, 1-5.8.6, 1-5.8.7.1, and Tables 3-5, 3-6, and 3-7, will not be subject to this requirement, and;

except a noninterfering shunt circuit, provided that a fault circuit condition on the shunt circuit wiring results only in the loss of the noninterfering feature of operation, and;

except connections to and between supplementary system components, provided that single open, ground, or short circuit conditions of the supplementary equipment or interconnecting means, or both, do not affect the required operation of the fire alarm system, and;

except the circuit of an alarm notification appliance installed in the same room with the central control equipment, provided that the notification appliance circuit conductors are installed in conduit or equivalently protected against mechanical injury, and;

except a trouble signal circuit, and;

except interconnection between equipment within a common enclosure, and;

except interconnection between enclosures containing control equipment located within 20 feet of each other where the conductors are installed in conduit or equivalently protected against mechanical injury, and;

except conductors for ground detection where a single ground does not prevent the required normal operation of the system, and;

except central station circuits serving notification appliances within a central station, and;

except pneumatic rate-of-rise systems of the continuous line type in which the wiring terminals of such devices are connected in multiple across electrically supervised circuits, and;

except interconnecting wiring of a stationary computer and the computer's keyboard, video monitor, mouse-type device, or touch screen, so long as the interconnecting wiring does not exceed 8 feet in length; is a listed computer/data processing cable as permitted by NFPA 70; and failure of cable does not cause the failure of the required system functions not initiated from the keyboard, mouse, or touch screen, and;

except communications and transmission channels extending from a supervising station to a subsidiary station(s) or protected premises, or both, which are compliant with the requirements of NFPA 72-1999 Chapter 5 and electrically isolated from the fire alarm system (or circuits) by a transmitter(s), are not required to monitor the integrity of installation conductors for a ground-fault condition, provided that a single ground condition does not affect the required operation of the fire alarm system.

NFPA 72-1999 1-5.8.6.1

If speakers are used to produce audible fire alarm signals, failure of any audio amplifier will result in an audible trouble signal and failure of any tone-generating equipment will result in an audible trouble signal. Tone generating and amplifying equipment enclosed as integral parts and serving only a single, listed loudspeaker will not be required to be monitored.

NFPA 72-1999 1-5.8.6.2

Where a two-way telephone communications circuit is used, its installation wires will be monitored for a short circuit fault that would cause the telephone communications circuit to become inoperative.

NFPA 72-1999 1-5.8.7.1

All primary and secondary power supplies will be monitored for the presence of voltage at the point of connection to the system. Failure of either supply will result in a trouble signal in accordance with NFPA 72-1999 1-5.4.6. The trouble signal also will be visually and audibly indicated at the protected premises. Where the DACT is power from a protected premises fire alarm system control unit, power failure indication will be in accordance with this paragraph, except a power supply for supplementary equipment and;

except the neutral of a three-, four-, or five-wire AC or DC supply source, and;

except in a central station, the main power supply, provided the fault condition is otherwise indicated so as to be obvious to the operator on duty, and;

except the output of an engine-driven generator that is part of the secondary power supply provided the generator is tested weekly in accordance with NFPA 72-1999 Chapter 7.

<table>
<tr><td>NEMA 2.17</td><td>

SUPPLEMENTARY CIRCUITS. Supplementary circuits may include of the following:

   a. Annunciators

   b. Signal sounding appliances not required by NFPA 72.

   c. Printers

   d. Fan shutdown

   e. Motor stop

   f. Elevator capture

These circuits may be unmonitored, provided the equipment to which they are connected is not required by NFPA 72, they are designated as supplementary by the authority having jurisdiction, and a short circuit, a break, or a ground fault does not prevent the required operation of the fire alarm system.

End-of-line resistors usually supervise circuits, however, end-of-line relays, diodes, and capacitors are also used. Supervision of devices and wiring in some addressable and multiplex systems is accomplished by polling wherein the control panel polls the devices and wiring then listens for the expected response.

</td></tr>
</table>

<table>
<tr><td>

NFPA 101-97 7-7.2.1

NFPA 101-97 A-7-7.2.1

</td><td>

Where supervised, automatic sprinkler systems are required by another section of NFPA 101-1997, supervisory attachments will be installed and monitored for integrity in accordance with NPFA 72, and a distinctive supervisory signal will be provided to indicate a condition that would impair the satisfactory operation of the sprinkler system. This will include, but not be limited to, monitoring of control valves, fire pump power supplies and running conditions, water tank levels and temperatures, pressure of tanks, and air pressure on dry-pipe valves. Supervisory signals will sound and will be displayed either at a location within the protected building that is constantly attended by qualified personnel or at an approved, remotely located receiving facility.

NFPA 72 gives details of standard practice in sprinkler supervision.

Subject to the approval of the AJH, sprinkler supervision may also be provided by direct connection to municipal fire departments or, in the case of very large establishments, to a private headquarters providing similar functions.

NFPA 72 covers such matters. Where municipal fire alarm systems are involved, reference should also be made to NFPA 1221.

</td></tr>
</table>

33006 DETECTION METHODS

REFERENCES:

NFPA 72

NFPA 101

DESCRIPTION:

Know the basic principles of automatic fire detectors listed in NFPA 72. Select the best type of detection for the application and ambient conditions.

Review element 33022, 33023, and 33024 for detector operation and guidelines.

NFPA 72-1999 1-4

NFPA 72-1999 A-1-4

NFPA 101

Cloud Chamber Smoke Detection. The principle of using an air sample drawn from the protected area into a high humidity chamber combined with a lowering of chamber pressure to create an environment in which the resultant moisture in the air condenses on any smoke particles present, forming a cloud, The cloud density is measured by a photoelectric principle. The density signal is processed and used to convey an alarm condition when it meets pre-set criteria.

Ember. A particle of solid material that emits radiant energy due either to its temperature or the process of combustion on its surface. Class A and Class D combustibles burn as embers under conditions where the flame typically associated with fire does not necessarily exist. This glowing combustion yields radiant emissions in parts of the radiant energy spectrum that are radically different from those parts affected by flaming combustion. Specialized detectors, specifically designed to detect those emissions, should be used in applications where this type of combustion is expected. In general, flame detectors are not intended for the detection of embers.

Fixed Temperature Detector. A device that responds when its operating element becomes heated to a predetermined level. The difference between the operating temperature of a fixed temperature device and the surrounding air temperature is proportional to the rate at which the temperature is rising and is commonly referred to as "thermal lag." The air temperature is always higher than the operating temperature of the device.

Flame. A body or stream of gaseous material involved in the combustion process and emitting radiant energy at specific wavelength bands determined by the combustion chemistry of the fuel. In most cases, some portion of the emitted radiant energy is visible to the human eye.

Flame Detector. A radiant energy-sensing fire detector that detects the radiant energy emitted by a flame (*See A-5-4.2.*)
NOTE: Flame detectors are categorized as ultraviolet, single wavelength infrared, ultraviolet infrared, or multiple wavelength infrared.

Heat Detector. A fire detector that senses heat produced by burning substances. Heat is the energy produced by combustion that causes substances to rise in temperature.

Ionization Smoke Detection. The principle of using a small amount of radioactive material to ionize the air between two differentially charged electrodes to sense the presence of smoke particles. Smoke particles entering the ionization volume decrease the conductance of the air by reducing ion mobility. The reduced conductance signal is processed and used to convey an alarm condition when it meets pre-set criteria. Ionization smoke detection is more responsive to invisible particles (smaller than 1 micron in size) produced by most flaming fires. It is somewhat less responsive to the larger particles typical of most smoldering fire. Smoke detectors utilizing the ionization principle are usually of the spot type.

Photoelectric Light Obscuration Smoke Detection. The response of photoelectric light obscuration smoke detectors is usually not affected by the color of smoke. Smoke detectors utilizing the light obscuration principle are usually of the line type. These detectors are commonly refered to as "projected beam smoke detectors."

Photoelectric Light-Scattering Smoke Detection. Photoelectric light-scattering smoke detection is more responsive to visible particles (larger than 1 micron in size) produced by most smoldering fires. It is somewhat less responsive to the smaller particles typical of most flaming fires.

Rate Compensation Detector. A device that responds when the temperature of the air surrounding the device reaches a predetermined level, regardless of the rate of temperature rise. A typical example is a spot-type detector with a tubular casing of a metal that tends to expand lengthwise as it is heated and an associated contact mechanism that closes at a certain point in the elongation. A second metallic element inside the tube exerts an opposing force on the contacts, tending to hold them open. The forces are balanced in such a way that, on a slow rate-of-temperature rise, there is more time for heat to penetrate to the inner element, which inhibits contact closure until the total device has been heated to its rated temperature level. However, on a fast rate-of-temperature rise, there is not as much time for heat to penetrate to the inner element, which exerts less of an inhibiting effect so that contact closure is achieved when the total device has been heated to a lower temperature. This, in effect, compensates for thermal lag.

Rate-of-Rise Detector. A device that responds when the temperature rises at a rate exceeding a predetermined value.

Smoke Detector. A device that detects visible or invisible particles of combustion.

Spark. A moving ember. The overwhelming majority of applications involving the detection of Class A and Class D combustibles with radiant energy-sensing detectors involve the transport of particulate solid materials through pneumatic conveyor ducts or mechanical conveyors. It is common in the industries that include such hazards to refer to a moving piece of burning material as a "spark" and to systems for the detection of such fires as "spark detection systems".

Wavelength. The distance between the peaks of a sinusoidal wave. All radiant energy can be described as a wave having a wavelength. Wavelength serves as the unit of measure for distinguishing between different parts of the spectrum. Wavelengths are measured in microns (:M), or Angstroms (Å).

Wavelength. The concept of wavelength is extremely important in selecting the proper detector for a particular application. There is a precise interrelation between the wavelength of light being emitted from a flame and the combustion chemistry producing the flame. Specific subatomic, atomic, and molecular events yield radiant energy of specific wavelengths. For example, ultraviolet photons are emitted as the result of the complete loss of electrons or very large changes in electron energy levels. During combustion, molecules are violently torn apart by the chemical reactivity of oxygen, and electrons are released in the process, recombining at drastically lower energy levels, thus giving rise to ultraviolet rations. Visible radiation is generally the result of smaller changes in electron energy levels within the molecules of fuel, flame intermediates, and products of combustion. Infrared radiation comes from the vibration of molecules or parts of molecules when they are in the superheated state associated with combustion. Each chemical compound exhibits a group of wavelengths at which it is resonant. These wavelengths constitute the chemical's infrared spectrum, which is usually unique to that chemical.

This interrelationship between wavelength and combustion chemistry affects the relative performance of various types of detectors with respect to various fires.

NFPA 72-1999 2-1.3.3

Initiating devices will be installed in all areas required by other NFPA codes and standards or by the AHJ. Each installed initiating device will be accessible for periodic maintenance and testing.

NFPA 72-1999 2-1.4.2.1

If required, total coverage will include all rooms, halls, storage areas, basements, attics, spaces above suspended ceilings, and other subdivisions and accessible spaces; and the inside of all closets, elevator shafts, enclosed stairways, dumbwaiter shafts, and chutes. Inaccessible areas will not be required to be protected by detectors

except where inaccessible areas contain combustible material, they will be made accessible and will be protected by a detector(s), and;

except detectors will not be required in combustible blind spaces where any of the following conditions exist:

    (a) Where the ceiling is attached directly to the underside of the supporting beams of a combustible roof or floor deck;

    (b) Where the concealed space is entirely filled with a noncombustible insulation (In solid joist construction, the insulation will be required to fill only the space from the ceiling to the bottom edge of the joist of the roof or floor deck.);

    (c) Where there are small, concealed spaces over rooms, provided any space in question does not exceed 50 ft^2 in area;

    (d) In spaces formed by sets of facing studs or solid joists in walls, floors, or ceilings where the distance between the facing studs or solid joists is less than 6 inches, and;

except detectors will not be required below open grid ceilings where all of the following conditions exist:

    (a) The openings of the grid are 1/4 inch or larger in the smallest dimension;

    (b) The thickness of the material does not exceed the smallest dimension;

    (c) The openings constitute at least 70 percent of the area of the ceiling material;

and;

except concealed, accessible spaces above suspended ceilings, used as a return air plenum meeting the requirements of NFPA 90A, where equipped with smoke detection at each connection from the plenum to the central air-handling system and;

except detectors will not be required underneath open loading docks or platforms and their covers and for accessible underfloor spaces if all of the following conditions exist;

> (a) Space is not accessible for storage purposes or entrance of unauthorized persons and is protected against the accumulations of windborne debris;
>
> (b) Space contains no equipment such as steam pipes, electric wiring, shafting, or conveyors;
>
> (c) Floor over the space is tight;
>
> (d) No flammable liquids are processed, handled, or stored on the floor above.

NFPA 72-1999 2-3.6.1 The selection and placement of smoke detectors will take into account both the performance characteristics of the detector and the areas into which the detectors are to be installed to prevent nuisance alarms or improper operation after installation. NFPA 72-1999 2-3.6.1.1 through 2-3.6.1.3 will apply.

NFPA 72-1999 2-3.6.1.2 The location of smoke detectors will be based on an evaluation of potential ambient sources of smoke, moisture, dust, or fumes, and electrical or mechanical influences to minimized nuisance alarms.

NFPA 72-1999 A-2-3.6.1.2 Smoke detectors can be affected by electrical and mechanical influences and by aerosols and particulate matter found in protected spaces. The location of detectors should be such that the influences of aerosols and particulate matter from sources such as those in NFPA 72-1999 Table A-2-3.6.1.2(a) are minimized. Similarly, the influences of electrical and mechanical factors shown in NFPA 72-1999 Table A-2-3.6.1.2(b) should be minimized. While it might not be possible to isolate environmental factors totally, an awareness of these factors during system layout and design favorably affects detector performance.

NFPA 72-1999 2-3.6.1.4 The effect of stratification below the ceiling will be considered. The guidelines in Appendix B will be permitted to be used.

NFPA 72-1999 2-3.1.1

The purpose of NFPA 72-1999 2-3 is to provide information to assist in design and installation of reliable early warning smoke detection systems for protection of life and property.

NFPA 72-1999 2-3.2

Smoke detectors will be installed in all areas where required either by the applicable laws, codes, or standards.

NFPA 72-1999 A-2-3.2

The person designing an installation should keep in mind that, in order for a smoke detector to respond, the smoke has to travel from the point of origin to the detector. In evaluating any particular building or location, likely fire locations should be determined first. From each of these points of origin, paths of smoke travel should be determined. Wherever practical, actual field tests should be conducted. The most desired locations for smoke detectors are the common points of intersection of smoke travel from fire locations throughout the building.

NOTE: This is one of the reasons that specific spacing is not assigned to smoke detectors by the testing laboratories.

NFPA 72-1999 A-2-3.4.1

For operation, all types of smoke detectors depend on smoke entering the sensing chamber or light beam. Where sufficient concentration is present, operation is obtained. Since the detectors are usually mounted on the ceiling, response time depends on the nature of the fire. A hot fire rapidly drives the smoke up to the ceiling. A smoldering fire, such as in a sofa, produces little heat; therefore, the time for smoke to reach the detector is increased.

NFPA 72-1999 A-2-3.4.3

In high ceiling areas, such as atriums, where spot-type smoke detectors are not accessible for periodic maintenance and testing, projected beam-type or air sampling-type detectors should be considered where access could be provided.

NFPA 72-1999 2-3.6.5

Where smoke detectors are installed to actuate a suppression system, NFPA 13 will apply.

NFPA 72-1999 A-2-3.6.5

For the most effective detection of fire in high rack storage areas, detectors should be located on the ceiling above each aisle and at intermediate levels in the racks. This is necessary to detect smoke that is trapped in the racks at an early stage of fire development, when insufficient thermal energy is released to carry the smoke to the ceiling. Earliest detection of smoke is achieved by locating the intermediate level detectors adjacent to alternate pallet sections as shown in Figures A-2-3.6.5 (a) and A-2-3.6.5 (b). The detector manufacturer's recommendations and engineering judgment should be followed for specific installations. A projected beam-type detector may be permitted to be used in lieu of a single row of individual spot-type smoke detectors. Sampling ports of an air sampling-type detector may be permitted to be located above each aisle to provide coverage equivalent to the location of spot-type detectors. The manufacturer's recommendations and engineering judgment should be followed for specific installations.

A projected beam-type detector is permitted to be used in lieu of a single row of individual spot-type smoke detectors.

Sampling ports of an air sampling-type detector can be permitted to be located above each aisle to provide coverage that is equivalent to the location of spot-type detectors. The manufacturer's recommendations and engineering judgment should be followed for the specific installation.

NFPA 72-1999 2-4.2.1 — The type and quantity of radiant energy-sensing fire detectors will be determined based on the performance characteristics of the detector and an analysis of the hazard, including the burning characteristics of the fuel, the fire growth rate, the environment, the ambient conditions, and the capabilities of the extinguishing media and equipment.

NFPA 72-1999 A-2-4.2.1 — The radiant energy emitted from any spark/ember is comprised of emissions in various bands of the ultraviolet, visible, and infrared portions of the spectrum. The relative quantities of radiation emitted in each part of the spectrum are determined by the fuel chemistry, the temperature, and the rate of combustion. The detector should be matched to the characteristics of the fire.

Almost all materials that participate in flaming combustion emit ultraviolet radiation to some degree during flaming combustion, whereas only carbon-containing fuels emit significant radiation at the 4.35 micron (carbon dioxide) band used by many detector types to detect a flame.

The radiant energy emitted from an ember is determined primarily by the fuel temperature (Planck's Law Emissions) and the emissivity of the fuel. Radiant energy from an ember is primarily infrared and, to a lesser degree, visible in wavelength. In general, embers do not emit ultraviolet energy in significant quantities (0.1 percent of total emissions) until the ember achieves temperatures of 2000°K (1727°C or 3240°F). In most cases, the emissions are included in the band of 0.8 microns to 2.0 microns, corresponding to temperatures of approximately 750°F to 1830°F (398°C to 1000°C).

NFPA 72-1999 A-2-4.3.2.1 The following are types of applications for which flame detectors are suitable:

1. High-ceiling, open-spaced buildings such as warehouses and aircraft hangers;

2. Outdoor or semi-outdoor areas where winds or draughts can prevent smoke from reaching a heat or smoke detector;

3. Areas where rapidly developing flaming fires can occur, such as aircraft hangers, petrochemical production, storage, and transfer areas, natural gas installations, paint shops, or solvent areas;

4. Areas needing high fire risk machinery or installations, often coupled with an automatic gas extinguishing system;

5. Environments that are unsuitable for other types of detectors.

Some extraneous sources of radiant emissions that have been identified as interfering with the stability of flame detectors include:

1. Sunlight

2. Lightning

3. X-rays

4. Gamma rays

5. Cosmic rays

6. Ultraviolet radiation from arc welding

7. Electromagnetic interference (EMI, RFI)

8. Hot objects

9. Artificial lighting.

33007 DETECTOR SPACING

REFERENCES:

NFPA 72

UL Fire Protection Equipment Directory

DESCRIPTION:

Know the spacing and location rules for heat, smoke, and flame sensing automatic fire detectors, including spacing on smooth ceilings, in irregular shaped areas, high ceilings, ceilings with joists or beams and peaked ceilings. Determine the number of each type of detector to protect a given building area.

NFPA 72-1999 2-2

Heat-Sensing Fire Detectors. Heat detectors will be installed as required by either the appropriate NFPA standard or the AHJ.

NFPA 72-1999 A-2-2.1.2

The linear space rating is the maximum allowable distance between heat detectors. The linear space rating is also a measure of the heat detector response time to a standard test fire where tested at the same distance; the higher the rating, the faster the response time. NFPA 72-1999 only recognizes heat detectors with a linear space rating of 50 feet or more.

NFPA 72-1999 2-2.2.1

NFPA 72-1999 Figure A-2.2.2.1

Spot-type heat detectors are to be located outside the dead-air space, that is, on the ceiling not less than 4 inches from the sidewall or on the sidewall between 4 inches and 12 inches from the ceiling. In the case of solid open joist construction, detectors are to be mounted on the bottom of the joists and in the case of beam construction where beams are less than 12 inches deep and less than 8 feet on center, detectors are permitted to be installed on the bottom of beams.

NFPA 72-1999 2-2.2.2

Line-type heat detectors are to be located on the ceiling or on the sidewall not more that 20 inches from the ceiling. In the case of solid open joist construction, detectors are to be mounted at the bottom of the joist and in the case of beam construction where beams are less than 12 inches deep and less than 8 feet on center, detectors are permitted to be installed on the bottom of beams. Where a line-type detector is used in an application other than open area protection, the manufacturer's installation instructions will be followed.

NFPA 72-1999 2-2.3

NFPA 72-1999A-2-2.3

Temperature. Heat detectors having fixed-temperature or rate-compensated elements are to be selected in accordance with Table 2-2.1.1.1 for the maximum ceiling temperature that can be expected.

NFPA 72-1999 A-2.2.4

Bear in mind that in addition to the special requirements for heat detectors installed on ceilings with exposed joists, reduced spacing may also be required due to other structural characteristics of the protected area or possible drafts or other conditions that could affect detector operation.

NFPA 72-1999 2-2.4.1.1

Smooth Ceiling Spacing of Heat Detectors.

- Either the distance between heat detectors will not exceed their listed spacing and there will be heat detectors within a distance of 1/2 listed spacing, measured at a right angle, from all walls or partitions extending to within 18 inches of the ceiling; or

- All points on the ceiling will have a heat detector within a distance equal to 0.7 times the listed spacing (0.7S). This is useful in calculating locations in corridors or irregular areas.

NFPA 72-1999 2-2.4.5.1

High Ceilings. On ceilings 10 feet to 30 feet high, heat detector linear spacing will be reduced in accordance with Table 2-2.4.5.1. Table 5-2.4.1.2 does not apply to the following detectors, which rely on the integration effect:

(a) Line-type electrical conductivity detectors [See NFPA 72-1999 A-1-4];

(b) Pneumatic rate-of-rise tubing [See NFPA 72-1999 A-1-4];

(c) Series connected thermoelectric effect detectors [See NFPA 72-1999 A-1-4].

In these cases, the manufacturer's recommendations will be followed for appropriate alarm point and spacing.

NFPA 72-1999 2-2.4.2

Solid Joist Construction. The spacing of heat detectors where measured at right angles to the solid joists, will not exceed 50 percent of the smooth ceiling spacing permitted under NFPA 72-1999 2-2.4.1 and 2-2.4.1.1.

NFPA 72-1999 2-2.4.3

Beam Construction. A ceiling will be treated as a smooth ceiling where the beams project no more than 4 inches below the ceiling. The spacing of spot-type heat detectors at right angles to the direction of beam travel will be not more than 2/3 of the smooth ceiling spacing permitted in accordance with NFPA 72-1999 2-2.4.1.1 and 2-2.4.1.2. Where the beams project more than 18 inches below the ceiling and are more than 8 feet on center, each bay formed by the beams will be treated as a separate area.

NFPA 72-1999 2-2.4.4.1

NFPA 72-1999
Figure A-2-2.4.4.1

Peaked Sloping Ceilings. A row of heat detectors must first be spaced and located at or within 3 feet of the peak of the ceiling, measured horizontally. The number and spacing of additional detectors, if any, will be based on the horizontal projection of the ceiling in accordance with the type of ceiling construction.

NFPA 72-1999 2-2.4.4.2

NFPA 72-1999
Figure A-2-2.4.4.2

Shed Sloping Ceilings. Sloped ceilings having a rise greater than 1 foot in 8 feet will have a row of detectors located on the ceiling within 3 feet of the high side of the ceiling measured horizontally, spaced in accordance with the type of construction. The remaining detectors, if any, will be located in the remaining area on the basis of the horizontal projection of the ceiling.

NFPA 72-1999 2-2.4.4.3

For a roof slope of less than 30 degrees, all detectors will be spaced utilizing the height at the peak. For a roof slope of greater than 30 degrees, the average slope height will be used for all detectors other than those located in the peak.

NFPA 72-1999 2-2.4.4.3

For a roof slope of less than 30 degrees, all detectors willl be spaced using the height at the peak. For a roof slope of greater than 30 degrees, the average slope height will be used for all detectors other than those located in the peak.

NFPA 72-1999 2-3.4.1.1

The location and spacing of smoke detectors are as a result of an evaluation based upon the guidelines detailed in NFPA 72-1999 and on engineering judgment. Some of the conditions to be considered include:

(a) Ceiling shape and surface;

(b) Ceiling height;

(c) Configuration of contents in the area to be protected;

(d) Burning characteristics of the combustible materials present;

(e) Ventilation;

(f) Ambient environment.

NFPA 72-1999 2-3.4.1.2 Where the intent is to protect against a specific hazard, the smoke detector(s) is(are) permitted to be installed closer to the hazard in a position where the detector is more likely to intercept the smoke.

NFPA 72-1999 2-3.4.5.1.1 On smooth ceilings, spacing of 30 feet is permitted to be used as a guide. In all cases, the manufacturer's documented instructions will be followed. Other spacing is permitted to be used depending upon ceiling height, different conditions, or response requirements. For the detection of flaming fires see NFPA 72-1999 Appendix B.

NFPA 72-1999 2-3.4.5.1.2 For smooth ceilings, all points on the ceiling will have a smoke detector within a distance equal to 0.7 times selected spacing.

NFPA 72-1999 2-3.4.6 **Solid Joist and Beam Construction.** Solid joists over 1 foot in depth are to be considered equivalent to beams for smoke detector spacing guidelines.

NFPA 72-1999 2-3.4.6.1 **Flat Ceilings.**
For ceiling heights of 12 feet or lower and beams or solid joist depths of 1 foot or less, smooth ceiling spacing running in the direction parallel to the beams or solid joists will be used and 1/2 the smooth ceiling spacing will be used in the direction perpendicular to the run of the beams or solid joists. For beam depths over 1 foot, spot-type smoke detectors will be permitted to be located either on the ceiling or on the bottom of the beams.

For beam depths exceeding 1 foot or for ceiling heights exceeding 12 feet, spot-type smoke detectors will be located on the ceiling in every beam pocket.

For solid joists, smoke detectors are to be located on the bottom of the joists.

NFPA 72-1999 2-3.4.6.2

Sloped Ceilings.
For beamed ceilings with beams running parallel to or up the slope, the spacing for flat beamed ceilings is to be used. The ceiling height is to be taken as the average height over the slope. For slopes greater than 10 degrees, smoke detectors located at 1/2 the spacing from the low end are not required. Spacing is to be measured along a horizontal projection of the ceilings.

For beamed ceilings with beams running perpendicular to or across the slope, spacing for flat beamed ceilings is to be used. The ceiling height will be taken as the average height over the slope.

For solid joists, smoke detectors are to be located on the bottom of the joists.

NFPA 72-1999 2-3.4.7

*NFPA 72-1999
Figure A-2-2.4.4.1*

Peaked Ceilings. Smoke detectors are to be spaced and located within 3 feet of the peak, measured horizontally. The number and spacing of additional detectors, if any, is to be based upon the horizontal projection of the ceiling.

NFPA 72-1999 2-3.4.8

*NFPA 72-1999
Figure A-2-2.4.4.2*

Shed Ceilings. Smoke detectors are to be spaced and located within 3 feet of the high side of the ceiling, measured horizontally. The number and spacing of additional smoke detectors, if any, is based upon the horizontal projection of the ceiling.

NFPA 72-1999 2-3.5.2.1

NFPA 72-1999 2-3.6.1.1

In under-floor and above-ceiling spaces that are used as HVAC plenums, smoke detectors are to be listed for the anticipated environment. Smoke detector spacing and locations are to be selected based upon anticipated airflow patterns and fire type.

NFPA 72-1999 2-3.6.6.2

Location. Smoke detectors are not to be located directly in the airstream of supply registers.

NFPA 72-1999 2-3.5.1

NFPA 72-1999 A-2-3.5.1

In spaces served by air-handling systems, detectors are not to be located where air from supply diffusers could dilute smoke before it reaches the detectors.

Detectors should not be located in a direct airflow nor closer than 3 feet from an air supply diffuser or return air opening.

NFPA 72-1999 2-3.6.5

NFPA 72-1999 A-2-3.6.5

NFPA 72-1999
Figure A-2-3.6.5 (a) & (b)

High Rack Storage. Where smoke detectors are installed to actuate a suppression system, NFPA 13 applies. For the most effective detection of fire in high rack storage areas, detectors should be located on the ceiling above each aisle and at intermediate levels in the racks. This is necessary to detect smoke that is trapped in the racks at an early stage of fire development, when insufficient thermal energy is released to carry the smoke to the ceiling. Earliest detection of smoke is achieved by locating the intermediate level detectors adjacent to alternate pallet sections. The detector manufacturer's recommendations and engineering judgment should be followed for specific installations.

A projected beam-type detector is permitted to be used in lieu of a single row of individual spot-type smoke detectors.

Sampling ports of an air sampling-type detector are permitted to be located above each aisle to provide coverage equivalent to the location of spot-type detectors. The manufacturer's recommendations and engineering judgement should be followed for the specific installation.

NFPA 72-1999 2-4.3.1

Spacing Considerations for Flame Detectors. The location and spacing of flame detectors will be the result of an engineering evaluation that takes into consideration:

(a) The size of the fire that is to be detected

(b) The fuel involved

(c) The sensitivity of the detector

(d) The field of view of the detector

(e) The distance between the fire and the detector

(f) The radiant energy absorption of the atmosphere

(g) The presence of extraneous sources of radiant emissions

(h) The purpose of the detection system

(i) The response time required

NFPA 72-1999 A-2-4.3.2.1

Flame Detector Applications and Stability. The types of application for which flame detectors are suitable are:

- High-ceiling, open-spaced buildings such as warehouses and aircraft hangers;

- Outdoor or semi outdoor areas where winds or draughts can prevent smoke from reaching a heat or smoke detector;

- Areas where rapidly developing flaming fires can occur, such as aircraft hangers, petrochemical production, storage, and transfer areas, natural gas installations, paint shops, or solvent areas;

- Areas needing high fire risk machinery or installations, often coupled with an automatic gas extinguishing system;

- Environment that is unsuitable for other types of detectors.

Some extraneous sources of radiant emissions that have been identified as interfering with the stability of flame detectors include:

- Sunlight

- Lightning

- X-rays

- Gamma rays

- Cosmic rays

- Ultraviolet radiation from arc welding

- Electromagnetic interference (EMI, RFI)

- Hot objects

- Artificial lighting.

Notes

33008 POWER SUPPLIES

REFERENCES:

NFPA 70
NFPA 72
NFPA 101
NFPA 110

Fire Alarm
Signaling Systems

DESCRIPTION:

Know the required primary (main), secondary (standby) and trouble power supply sources for each type of fire alarm system. Calculate the size storage battery needed to supply a given system for various standby times.

NFPA 72-1999 1-5.2.3

Power Supplies. Fire alarm systems will be provided with at least two independent and reliable power supplies, one primary and one secondary (standby), each of which will be of adequate capacity for the application.

Where the primary power is supplied by a dedicated branch circuit of an emergency system in accordance with NFPA 70, National Electrical Code, Article 700, or a legally required standby system in accordance with NFPA 70, National electrical Code, Article 701, a secondary supply is not required.

Where the primary power is supplied by a dedicated branch circuit of an optional standby system in accordance with NFPA 70, National Electrical Code, Article 702, which also meets the performance requirements of Article 700 or Article 701, a secondary supply is not required.

Where DC voltages are employed, they are to be limited to no more than 350 volts above earth ground.

NFPA 72-1999 1-5.2.4

Primary Supply. The primary supply will have a high degree of reliability, will have adequate capacity for the intended service, and will consist of one of the following:

(a)   Light and power service arranged in accordance with 1-5.2.5;

(b)   Where a person specifically trained in its operation is on duty at all times, an engine-driven generator or equivalent arranged in accordance with 1-5.2.10.

NFPA 72-1999 1-5.2.5.1   A light and power service employed to operate the system under normal conditions will have a high degree of reliability and capacity for the intended service. This service will consist of one of the following:

> **(a) Two-Wire Supplies.** A two-wire supply circuit is permitted to be used for either the primary operating power supply or the trouble signal power supply of the signaling system.

> **(b) Three-Wire Supplies.** A three-wire AC or DC supply circuit having a continuous unfused neutral conductor, or a polyphase ac supply circuit having a continuous unfused neutral conductor where interruption of one phase does not prevent operation of the other phase, will be permitted to be used with one side or phase for the primary operating power supply and the other side or phase for the trouble signal power supply of the fire alarm system.

NFPA 72-1999 1-5.2.5.2   Connections to the light and power service is to be on a dedicated branch circuit(s). The circuit(s) and connections will be mechanically protected. Circuit disconnecting means will have a red marking, will be accessible only to authorized personnel, and will be identified as FIRE ALARM CIRCUIT CONTROL. The location of the circuit disconnecting means will be permanently identified at the fire alarm control unit.

NFPA 72-1999 1-5.2.5.3   **Overcurrent Protection.** An overcurrent protection device of suitable current-carrying capacity and capable of interrupting the maximum short-circuit current to which it may be subject is to be provided in each ungrounded conductor. The overcurrent protective device will be enclosed in a locked or sealed cabinet located immediately adjacent to the point of connection to the light and power conductors.

NFPA 72-1999 1-5.2.5.4   Circuit breakers or engine stops are not be installed in such a manner as to cut off the power for lighting or for operating elevators.

NFPA 72-1999 1-5.2.6   **Secondary Supply Capacity and Sources.** The secondary supply will automatically supply the energy to the system within 30 seconds, and without loss of signals, wherever the primary supply is incapable of providing the minimum voltage required for proper operation. The secondary (standby) power supply will supply energy to the system in the event of total failure of the primary (main) power supply or when the primary voltage drops to a level insufficient to maintain functionality of the control equipment and system components. Under maximum normal load, the secondary supply will have sufficient capacity to operate a protected premises, central station, or proprietary system for 24 hours, or an auxiliary or remote station system for 60 hours; and at the end of that period, will be capable of operating all alarm notification appliances used for evacuation or to direct aid to the location of an emergency for 5 minutes. The secondary power supply for emergency voice/alarm communications service will be capable of operating the system under maximum normal load for 24 hours and then will be capable of operating the system during a fire, or other emergency condition for a period of 2 hours. Fifteen minutes of evacuation alarm operation at maximum connected load will be considered the equivalent of 2 hours of emergency operation.

For a combination system, the secondary supply capacity required above will include the load of any non-fire related equipment, functions, or features which are not automatically disconnected upon transfer of operating power to the secondary supply.

The secondary supply will consist of one of the following:

> (a) A storage battery arranged in accordance with 1-5.2.9;
>
> (b) An automatic starting, engine-driven generator arranged in accordance with 1-5.2.10 and storage batteries with 4 hours of capacity arranged in accordance with 1-5.2.9;
>
> (c) Multiple engine-driven generators, one of which is arranged for automatic starting, arranged in accordance with 1-5.2.10, and capable of supplying the energy required herein, with the largest generator out of service. The second generator is permitted to be started by pushbutton.

Operation on secondary power will not affect the required performance of a fire alarm system. The system will produce the same alarm, supervisory, and trouble signals and indications (excluding the AC power indicator) when operating from the standby power source as are produced when the unit is operating form the primary power source.

NFPA 72-1999 1-5.2.7

Continuity of Power Supplies.
> (a) Where signals could be lost on transfer of power between the primary and secondary sources, rechargeable batteries of sufficient capacity to operate the system under maximum normal load for 15 minutes will assume the load in such a manner that no signals are lost where either of the following conditions exists:

(1) Secondary power is supplied in accordance with 1-5.2.6(a) or 1-5.2.6(b), and the transfer is made manually; or

(2) Secondary power is supplied in accordance with 1-5.2.6(c).

(a) Where signals will not be lost due to transfer of power between the primary and secondary sources, one of the following arrangements are to be made:

(1) The transfer will be automatic;

(2) Special provisions are to be made to allow manual transfer within 30 seconds of loss of power.

(3) The transfer is to be arranged in accordance with 1-5.2.6(a).

(a) Where a computer system of any kind or size is used to receive or process signals, an uninterruptible power supply (UPS) with sufficient capacity to operate the system for at least 15 minutes; or until the secondary supply is capable of supplying the UPS input power requirements, is required where either of the following conditions apply:

(1) The status of signals previously received will be lost upon loss of power;

(2) The computer system cannot be restored to full operation within 30 seconds of loss of power.

(a) A positive means for disconnecting the input and output of the UPS system while maintaining continuity of power supply to the load is to be provided.

NFPA 72-1999 1-5.2.8.2   Power supervisory devices are to be arranged so as not to impair the receipt of fire alarm or supervisory signals.

NFPA 72-1999 1-5.2.9.2.1   **Battery Charging.**
Adequate facilities are to be provided to automatically maintain the battery fully charged under all conditions of normal operation and, in addition, to recharge batteries within 48 hours after fully charged batteries have been subject to a single discharge cycle as specified in 1-5.2.5.3. Upon attaining a fully charged condition, the charge rate will not be so excessive as to result in battery damage.

NFPA 72-1999 1-5.2.9.2.2   Supervising stations are to maintain spare parts or units available, which will be used to restore failed charging capacity prior to the consumption of 1/2 of the capacity of the batteries for the supervising station equipment.

NFPA 72-1999 1-5.2.9.2.3   Batteries are to be either trickle-or float-charged.

NFPA 72-1999 1-5.2.9.2.4   A rectifier employed as a battery charging supply source is to be of adequate capacity. A rectifier employed as a charging means will be energized by an isolating transformer.

NFPA 72-1999 1-5.2.9.4   **Metering.** The charging equipment is to provide either integral meters or readily accessible terminal facilities for the connection of portable meters by which the battery voltage and charging current can be determined.

NFPA 72-1999 1-5.2.9.5   **Charger Supervision.** Supervision means appropriate for the batteries and charger employed will be provided to detect a failure of battery charging and initiate a trouble signal in accordance with 1-5.4.6.

NFPA 72-1999 1-5.2.10.2   **Engine-Driven Generator Capacity.** The unit is to be of a capacity that is sufficient to operate the system under the maximum normal load conditions in addition to all other demands placed upon the unit, such as those of emergency lighting.

NFPA 72-1999 1-5.2.10.3.1   **Engine-Driven Generator Fuel.** Fuel is to be stored in outside underground tanks wherever possible, and gravity feed is not to be used. If gasoline-driven generators are used, fuel will be supplied from a frequently replenished "working" tank, or other means provided, to ensure that the gasoline is always fresh because gasoline deteriorates with age.

NFPA 72-1999 1-5.2.10.3.2   Sufficient fuel will be available in storage for 6 months of testing plus the capacity specified in 1-5.2.5. For public fire alarm reporting systems, the requirements of Chapter 6 applies.

If a reliable source of supply is available at any time on 2-hour notice, sufficient fuel will be in storage for 12 hours of operation at full load.

Fuel systems using natural or manufactured gas supplied through reliable utility mains are not required to have fuel storage tanks unless located in seismic risk zone 3 or greater as defined in ANSI A-58.1, *Building Code Requirements for Minimum Design Loads in Buildings and Other Structures.*

NFPA 72-1999 1-5.2.10.4   **Battery and Charger.** A separate storage battery and separate automatic charger will be provided for starting the engine-driven generator and will not be used for any other purpose.

Notes

33009 SYSTEM ACCEPTANCE AND PERIODIC TESTS

REFERENCES:

NFPA 72

*Manufactures'
Test Manuals*

DESCRIPTION:

Understand the requirements and related procedures to conduct and document proper fire alarm system acceptance tests. Procedure includes (1) documentation of plans and acceptance test reports; (2) indicating approval authorities; (3) instructing owners as to the location of all alarm initiating, alarm indicating and all other related components of the system; (4) care and maintenance of the system; and (5) providing owner with appropriate technical data sheets and maintenance manuals issued by the equipment manufacturer. Know the periodic equipment and circuit testing procedures, including frequency of tests and method of testing each component and circuit of a fire alarm system.

NFPA 72-1999 7-1.2

The owner or owner's designated representative is responsible for inspection, testing, and maintenance of the system and alterations or additions to this system. The delegation of responsibility is to be in writing, with a copy of such delegation made available to the authority having jurisdiction upon request.

NFPA 72-1999 7-1.2.1

Inspection, testing, or maintenance is permitted to be done by a person or organization other than the owner where conducted under written contract. Testing and maintenance of central station service systems will be performed under the contractual arrangements specified in 5-2.2.2.

NFPA 72-1999 7-1.2.2

Service personnel will be qualified and experienced in the inspection, testing, and maintenance of fire alarm systems. Examples of qualified personnel is permitted to include, but not limited to, individuals with the following qualifications:

(a) Factory trained and certified;

(b) National Institute for Certification in Engineering Technologies fire alarm certified;

(c) International Municipal Signal Association fire alarm certified;

(d) Certified by a state or local authority;

(e) Trained and qualified personnel employed by an organization listed by a national testing laboratory for the servicing of fire alarm systems.

NFPA 72-1999 7-1.3.1

Before proceeding with any testing, all persons and facilities receiving alarm, supervisory, or trouble signals, and all building occupants, are to be notified of the testing to prevent unnecessary response. At the conclusion of testing, those previously notified (and others, as necessary) will be notified that testing has been concluded.

NFPA 72-1999 7-1.4

Prior to system maintenance or testing, the system certificate and the information regarding the system and system alterations, including specifications, wiring diagrams, and floor plans, will be provided by the owner or a designated representative to the service personnel upon request.

NFPA 72-1999 7-1.5.3

Discharge testing of suppression systems is not required by this code. Suppression systems will be secured from inadvertent actuation, including disconnection of releasing solenoids or electric actuators, closing of valves, other actions, or combinations thereof, as appropriate for the specific system, for the duration of the fire alarm system testing.

NFPA 72-1999 7-1.5.4

Testing will include verification that the releasing circuits and components energized or actuated by the fire alarm system are electrically supervised and operate as intended on alarm.

NFPA 72-1999 7-1.6.1

Initial Acceptance Testing. All new systems will be inspected and tested in accordance with the requirements of this NFPA 72-1999 Chapter 7.

NFPA 72-1999 7-1.6.2.1

Reacceptance Testing. Reacceptance testing will be performed after any of the following:

- Added or deleted system components;

- Any modification, repair, or adjustment to system hardware or wiring;

- Any change to site-specific software.

All components, circuits, systems operations, or site-specific software functions known to be affected by the change or identified by a means that indicates the system operational changes are to be 100 percent tested. In addition, 10 percent of initiating devices that are not directly affected by the change, up to a maximum of 50 devices, also will be tested and proper system operation verified. A revised record of completion in accordance with 1-6.2.1 will be prepared to reflect any changes.

NFPA 72-1999 7-1.6.2.2

Changes to all control units connected or controlled by the system executive software requires a 10 percent functional test of the system, including a test of at least one device on each input and output circuit to verify critical system functions such as notification appliances, control functions, and off-premises reporting.

NFPA 72-1999 7-2.1

Central Stations. At the request of the AHJ, the installation will be inspected at the request of the authority having jurisdiction for complete information regarding the system, including specifications, wiring diagrams, and floor plans that have been submitted for approval prior to installation of equipment and wiring.

Review **Table 7-2.2 Test Methods.**

Review **Table 7-3.1 Visual Inspection Frequencies.**

Review **Table 7-3.2 Testing Frequencies.**

NFPA 72-1999 7-3.2.1

Smoke detector sensitivity will be checked within 1 year after installation and every alternate year thereafter. After the second required calibration test, where sensitivity test indicate that the detector has remained within its listed and marked sensitivity range (or 4 percent obscuration light gray smoke, if not marked), the length of time between calibration tests is permitted to be extended to a maximum of 5 years. Where the frequency is extended, records of detector-caused nuisance alarms and subsequent trends of these alarms will be maintained. In zones or in areas where nuisance alarms show any increase over the previous year, calibration tests will be performed.

To ensure that each smoke detector is within its listed and marked sensitivity range, it will be tested using any of the following methods:

    (a) Calibrated test method;

(b) Manufacturer's calibrated sensitivity test instrument;

(c) Listed control equipment arranged for the purpose;

(d) Smoke detector/control unit arrangement whereby the detector causes a signal at the control unit where its sensitivity is outside its acceptable sensitivity range;

(e) Other calibrated sensitivity test method acceptable to the authority having jurisdiction.

Detectors found to have a measured sensitivity outside the listed and marked sensitivity range are to be cleaned and recalibrated or replaced.

Detectors listed as field adjustable are permitted to be either adjusted within the listed and marked sensitivity range and cleaned and recalibrated, or replaced.

This requirement does not apply to single station detectors referenced in 7-3.3 and Table 7-2.2.

The detector sensitivity will not be tested or measured using any device that administers an unmeasured concentration of smoke or other aerosol into the detector.

NFPA 72-1999 7-5.1

Permanent Records. After successful completion of acceptance tests approved by the authority having jurisdiction, a set of reproducible as-built installation drawings, operation and maintenance manuals, and a written sequence of operation will be provided to the building owner or the owner's designated representative. The owner will be responsible for maintaining these records for the life of system for examination by any authority having jurisdiction. Paper or electronic media is permitted.

NFPA 72-1999 7-5.2.1

Maintenance, Inspection, and Testing Records. Records will be retained until the next test and for 1 year thereafter.

NFPA 72-1999 7-5.2.2

A permanent record of all inspections, testing, and maintenance is to be provided that includes the following information regarding tests and all the applicable information requested in Figure 7-5.2.2:

(a) Date;

(b) Test frequency;

(c) Name of property;

(d) Address;

(e) Name of person performing inspection, maintenance, tests, or combination thereof, and affiliation, business address, and telephone number;

(f) Name, address, and representative of approving agency(ies);

(g) Designation of the detector(s) tested, for example, "Tests performed in accordance with Section _____;

(h) Functional test of detectors;

(i) Functional test of required sequence of operations;

(j) Check of all smoke detectors;

(k) Loop resistance for all fixed-temperature, line-type heat detectors;

(l) Other tests as required by equipment manufacturers;

(m) Other tests as required by the authority having jurisdiction;

(n) Signatures of tester and approved authority representative;

(o) Disposition of problems identified during test, for example, owner notified, problem corrected/successfully retested, device abandoned in place.

See NFPA 72-1999 Figure 7-5.2.2 Inspection and Testing Form.

NFPA 72-1999 7-5.3

For supervising station fire alarm systems, records pertaining to signals received at the supervising station that result from maintenance, inspection, and testing, is to be maintained for not less than 12 months. Upon request, a hard copy record will be provided to the authority having jurisdiction. Paper or electronic media will be permitted.

NFPA 72-1999 7-5.4

If the operation of a device, circuit, control panel function, or special hazard system interface is simulated, it will be noted on the certificate that the operation was simulated, and the certificate will indicate by whom it was simulated.

Notes

33010 CONSTRUCTION PLANS

REFERENCES:

See general construction texts and references.

DESCRIPTION:

Understand construction plan symbols and terminology. Determine uses and sizes of various building areas from plan information. Determine location of structural obstructions and other mechanical systems.

Know and understand the following structures and the way they are used for installation purposes:

- Framing Plans;
- Columns;
- Girders;
- Beams;
- Joists;

Concrete Construction:

- Pre-cast;
- Cast in place;
- Schedules;

Steel Construction:

- Component Designation;
- Methods of Joining;

Wood Construction:

- Schedules;
- Lamination;

Definitions:

Americans with Disabilities Act (ADA). A law that makes it illegal to discriminate against disabled persons in the area of employment, public and private transportation, and access to public and commercial buildings.

ANSI American National Standards Institute

ASTM American Society of Testing and Materials

Beam A structural member transversely supporting a load.

Bearing Wall A wall that supports any vertical load in addition to its own weight.

Concrete A mixture of cement, sand, and gravel with water.

Conduction The flow of heat through an object by transferring heat from one molecule to another.

Contract An agreement between a seller and purchaser. The title is withheld from the purchaser until all required payments to the seller have been completed.

Convection Refers to the transfer of heat by a moving fluid (liquid or gas).

Easement An area of a piece of property giving rights to another for the purpose of placing power lines, drains, and other specified users.

Ell An extension or wing of a building at right angles to the main section.

Fill Sand, gravel, or loose earth used to bring a sub-grade up to a desired level around a building.

Foundation The supporting portion of a structure below the first floor construction, or below grade, including the footings.

Girder A large or principal beam of wood or steel used to support concentrated loads at isolated points along its length.

I-Beam A steel beam with a cross section resembling the letter I. It is used for long spans as basement beams or over wide wall openings, such as a double garage door when wall and roof loads are imposed on the opening.

Joist A horizontal structural member that supports the floor or ceiling system.

On Center (OC) The measurement of spacing for studs, rafters, joists, and other framing members from the center of one member to the center of the next.

Pier A masonry pillar usually below a building to support the floor framing.

Plenum System A system of heating or air conditioning in which the air is forced through a chamber connected to distributing ducts.

Precast Concrete shapes that are made before being placed into a structure.

Rafter One of a series of structural members of a roof designed to support roof loads. The rafters of a flat roof are sometimes called roof joists.

Suspended Ceiling A ceiling system supported by hanging from the overhead structural framing.

Truss Structural members arranged and fastened in triangular units to form a rigid framework for support of loads over a long span.

BUILDING MATERIAL SYMBOLS

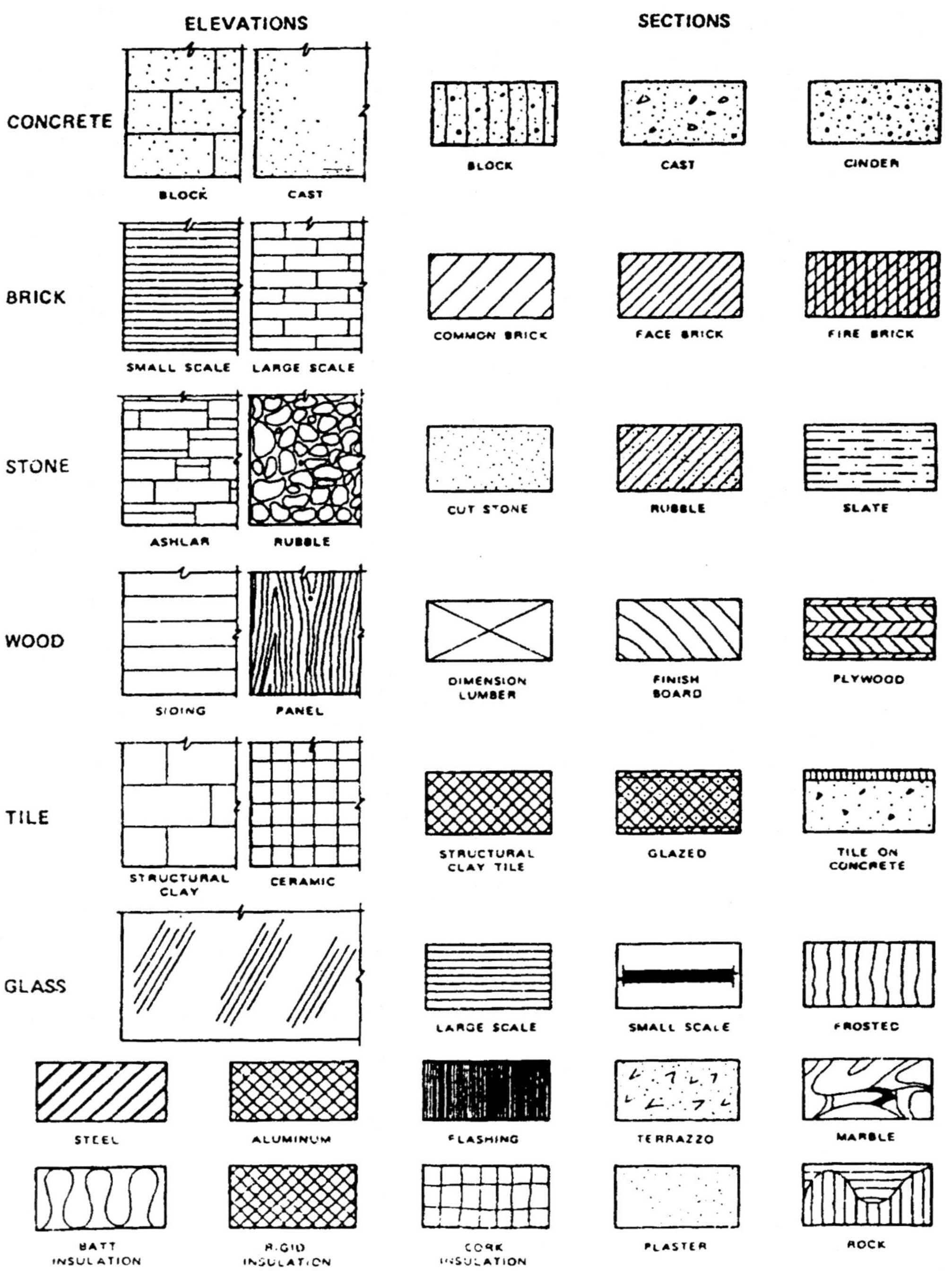

33011 SPECIFICATIONS AND COST ESTIMATES

REFERENCES:

AIA A201
NFPA 90A
NFPA 90B
NFPA 70

DESCRIPTION:

Read and interpret standard specs and become familiar with the existence of General Conditions, General Mechanical Conditions, and those specific fire alarm systems specifications pertaining to the fire alarm system. Distinguish those fire protection requirements often found in the heating, ventilating and air conditioning, or mechanical or electrical specifications. Be aware of the interface responsibility among fire detection devices, electrical interface requirements and alarm indicating appliances. Know cost estimate details as they relate to the preparation of working drawings.

CONSTRUCTION DOCUMENTS

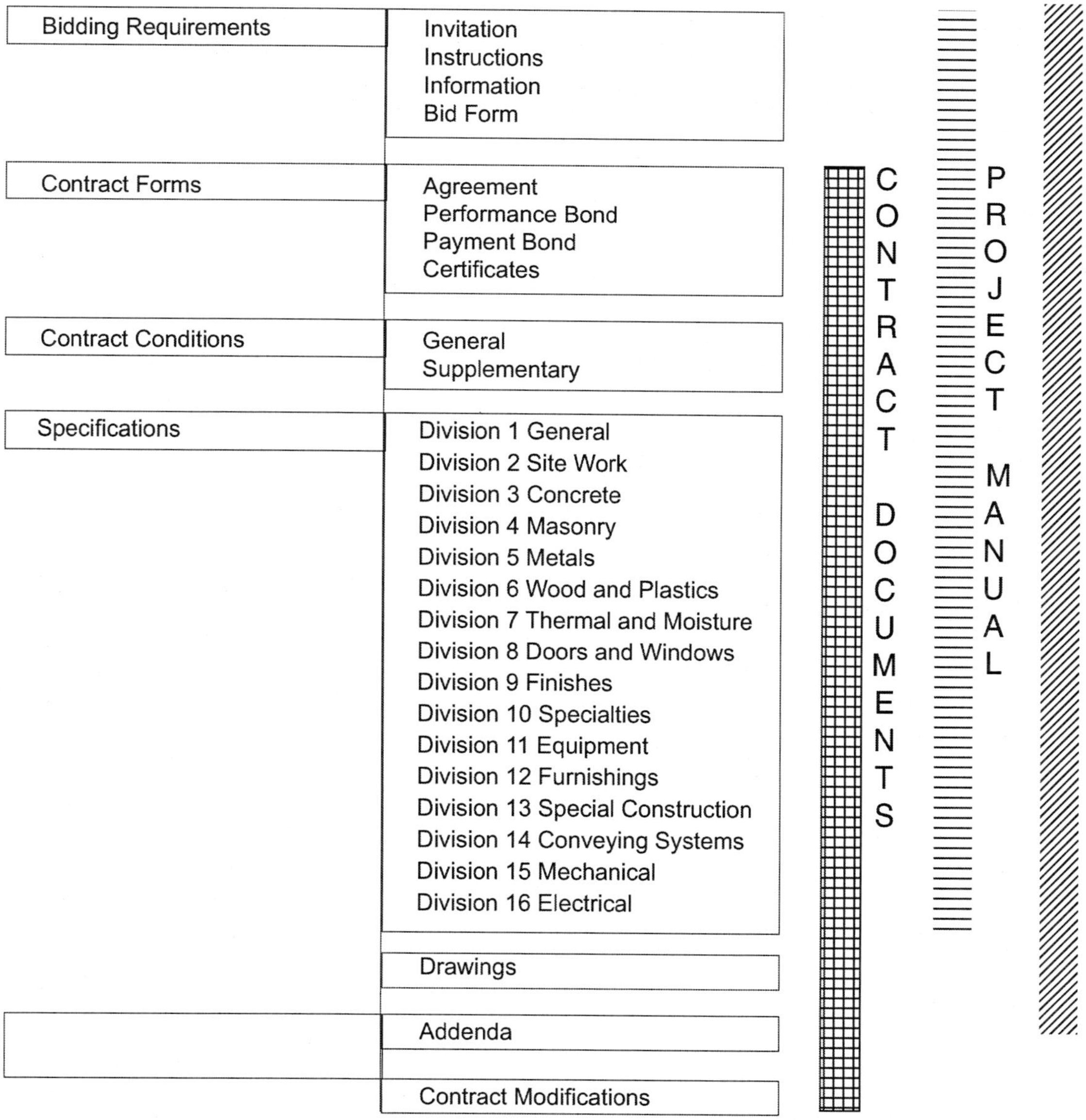

Bidding Requirements	Invitation Instructions Information Bid Form		
Contract Forms	Agreement Performance Bond Payment Bond Certificates		
Contract Conditions	General Supplementary		
Specifications	Division 1 General Division 2 Site Work Division 3 Concrete Division 4 Masonry Division 5 Metals Division 6 Wood and Plastics Division 7 Thermal and Moisture Division 8 Doors and Windows Division 9 Finishes Division 10 Specialties Division 11 Equipment Division 12 Furnishings Division 13 Special Construction Division 14 Conveying Systems Division 15 Mechanical Division 16 Electrical		
	Drawings		
	Addenda		
	Contract Modifications		

CONTRACT DOCUMENTS — PROJECT MANUAL — BIDDING DOCUMENTS

Specifications define the qualitative requirements for products, materials, and workmanship. Bidding requirements, forms and conditions of the contract are not specifications.

33012 CONTRACTS

REFERENCES:

Building Professional's Guide to Contract Documents

DESCRIPTION:

Understand contractual relationships in the construction industry.

Construction documents are defined as all of the written and graphic documents prepared or assembled by the architect/engineer for communicating the design and administering the construction contract.

Contract documents include:

- Bidding requirements;

- Contract forms;

- Conditions of the contract;

- Specifications;

- Drawings;

- Addenda;

- Contract modifications;

Contract definitions include:

Agreement. The written contract entered into by the owner and the contractor to perform work. The agreement is a legal instrument binding the parties to the work. The agreement defines the relationships and obligations that exist between the owner and the contractor.

Construction Performance Bond. Provides financial protection for the owner in the event the contractor does not complete the work in accordance with the agreement.

Construction Payment Bond. Protects the labor force and the suppliers of materials should the contractor fail to meet their obligations. Precludes the need for the labor force or the suppliers to seek payment directly from the owner because of non-payment by the contractor.

Certificates. Includes certificates of insurance and certificates of compliance with applicable laws and regulations.

General Conditions. General clauses that establish how the project is to be administered. General conditions contain provisions that are common practice nationwide. Standard documents published by professional societies are often used.

Specifications. Specifications define the qualitative requirements for products, materials, and workmanship upon which the contract is based.

Drawings. A graphic representation of the work to be done. Drawings illustrate the relationship of the materials to each other, including size, shape, location, and connections.

Addenda. Additions made to the construction documents during the bidding period. Addenda are used to revise, delete, or add to any of the bidding requirements or contract documents.

Contract Modifications. After the agreement has been signed, additions, deletions, or modifications to the work to be done are accomplished by change order, supplemental instruction, and/or field orders. Contract modifications may be issued at any time during the contract period.

33013 BUILDING CODES

REFERENCES:

NFPA 101

Basic/National Building Code

Standard Building Code

Uniform Building Code

Fire Protection Through Modern Building Codes

Fire Protection handbook

DESCRIPTION:

Understand building codes and their enforcement. Know the inter-relationship of building codes and installation standards. Review building codes to determine NFPA standards adoption and special requirements for fire alarm systems. Understand options of "trade-offs" available with regard to the elimination of sprinklers when automatic smoke detection systems are provided. For a particular building, compile information in tabular form that describes all trade-offs available under a model building code.

A building code is a law that sets forth minimum requirements for design and construction of buildings and structures. These minimum requirements are established to protect the health and safety of society. There are two types of building codes:

1) Specification codes spell out in detail what materials can be used, the building size, and how components should be assembled; and,

2) Performance codes detail the objective to be met and establish criteria for determining if the objective has been met.

The requirements contained in building codes are generally based upon the known properties of materials, the hazards presented by various occupancies, and on previous experiences.

A building code is law. Because of code complexity, many jurisdictions adopt model codes instead of writing their own.

Four independent groups have developed model codes:

1) American Insurance Association (AIA);

2) International Conference of Building Officials (ICBO);

3) Southern Building Code Congress International (SBCCI);

4) Building Officials and Code Administrators (BOCA).

Within the scope of the police power of state government is the regulation of building construction for the health and safety of the public, a power usually delegated to the local governments within the state.

The application of building code requirements usually applies to new construction and major alterations to a building. Building code application typically culminates in the issuance of an occupancy permit. After occupancy, fire prevention codes usually apply.

Many of the requirements found in building codes are excerpts from, or based upon, the standards published by nationally recognized organizations. Numerous NFPA standards are referenced by the model building codes and are given the status of statutory law when the standard is adopted by a state government.

33014 INSURANCE AUTHORITIES & THEIR REQUIREMENTS

REFERENCES:

DESCRIPTION:

Understand special interpretative guides, data sheets and other requirements of Factory Mutual System, Industrial Risk Insurers, Kemper Insurance Group, Insurance Services Office, etc. Prepare shop drawings for submittal to insurance authority.

Industrial Risk Insurers (IRI)

- Insurance underwriter

- Requires UL® listed equipment

- Requires UL® certification of system

- Hazardous and high risk coverage

- Sprinkler and suppression discharge

Factory Mutual (FM)
Consists of two groups: FMRC, research group, and FMI, insurance group.

> FMRC is the approval agency of FM that test equipment to FM standards and is accepted as an industry standard by other insurance underwriters.

> Allendale, Arkwright, and Protection Mutual Insurance own FMI. The objective of FMI is to help policyholders protect their properties.

Other insurance organizations include:

Insurance Service Organization (ISO)
- Rate setting

- Property insurers

- Sprinkler and watchman tours

American Insurance Agency (AIA)

Kemper Insurance Group
- Insurance underwriters

Notes

33015 GOVERNMENTAL AGENCIES

REFERENCES:

VA Specifications

*State, City, and
related documents*

DESCRIPTION:

Know special fire alarm installation requirements of government agencies such as DOD, VA, state, city, and others and their relationship to national installation standards. Prepare shop drawings for submittal to government authorities.

Veterans Administration (VA) requires photoelectric detectors, base loop reporting, and Buy American.

Government Services Administration (GSA) seven channel voice.

Department of Defense (DOD)

Department of Energy (DOE)

Army Corp of Engineers requires PE (Professional Engineer) stamp on drawings.

American Disabilities Act (ADA) applies to all buildings with public access and all areas within buildings which the public has access to. At this time, ADA requires no specific visible notification appliance distribution pattern. A single requirement of a minimum intensity of 75 candela in all areas, a minimum flash rate of 1 flash per second, wall mount only, andspacing such that no point in any room is more than 50 feet from a visible notification appliance.

California State Fire Marshal (CSFM)

New York City Bureau of Standards and Appeals (BSA)

North Carolina Department of Insurance requires alarm verification.

Notes

REFERENCES:

NFPA 72

NFPA 101

<u>*Fire Alarm*
Signaling Systems</u>

DESCRIPTION:

Understand a local fire alarm system, including its purpose, types of signaling service available, basic requirements, primary and secondary power supply sources, supervision, signal appliances and signal capacity of circuits.

NFPA 72-1999 1-4

Protected Premises (Local) Fire Alarm System. A protected premises system that sounds an alarm at the protected premises as the result of the manual operation of a fire alarm box or the operation of protection equipment or system, such as water flowing in a sprinkler system, the discharge of carbon dioxide, the detection of smoke, or the detection of heat.

NFPA 72-1999 1-5.1.2

Equipment. Equipment constructed and installed in conformity with this code is to be listed for the purpose for which it is used.

NFPA 72-1999 1-5.2.3

Power Sources. Fire alarm systems will be provided with at least two independent and reliable power supplies, one primary and one secondary (standby), each of which will be of adequate capacity for the application.

Where the primary power is supplied by a dedicated branch circuit of an emergency system in accordance with NFPA 70, National Electrical Code, Article 700, or a legally required standby system in accordance with NFPA 70, National electrical Code, Article 701, a secondary supply is not required.

Where the primary power is supplied by a dedicated branch circuit of an optional standby system in accordance with NFPA 70, National Electrical Code, Article 702, which also meets the performance requirements of Article 700 or Article 701, a secondary supply is not required.

Where DC voltages are employed, they will be limited to no more than 350 volts above earth ground.

NFPA 72-1999 1-5.2.4

Primary Supply. The primary supply will have a high degree of reliability, have adequate capacity for the intended service, and consist of one of the following:

1. Light and power service arranged in accordance with 1-5.2.5;

2. Where a person specifically trained in its operation is on duty at all times, an engine-driven generator or equivalent arranged in accordance with 1-5.2.10.

NFPA 72-1999 1-5.2.5.2

Connections to the light and power service will be on a dedicated branch circuit(s). The circuit(s) and connections will be mechanically protected. Circuit disconnecting means will have a red marking, be accessible only to authorized personnel, and be identified as FIRE ALARM CIRCUIT CONTROL. The overcurrent protective device will be enclosed in a locked or sealed cabinet located immediately adjacent to the point of connection to the light and power conductors.

NFPA 72-1999 1-5.2.6

Secondary Supply Capacity and Sources. The secondary supply will automatically supply the energy to the system within 30 seconds, and without loss of signals, wherever the primary supply is incapable of providing the minimum voltage required for proper operation. The secondary (standby) power supply will supply energy to the system in the event of total failure of the primary (main) power supply or when the primary voltage drops to a level insufficient to maintain functionality of the control equipment and system components. **Under maximum normal load, the secondary supply will have sufficient capacity to operate a protected premises, central station, or proprietary system for 24 hours, or an auxiliary or remote station system for 60 hours; and, at the end of that period, will be capable of operating all alarm notification appliances used for evacuation or to direct aid to the location of an emergency for 5 minutes.** The secondary power supply for emergency voice/alarm communications service will be capable of operating the system under maximum normal load for 24 hours and then be capable of operating the system during a fire, or other emergency condition for a period of 2 hours. Fifteen minutes of evacuation alarm operation at maximum connected load is considered the equivalent of 2 hours of emergency operation.

NFPA 72-1999 1-5.4.2.1

Coded Alarm Signals. A coded alarm will consist of not less than three complete rounds of the number transmitted. Each round will consist of not less than three impulses.

NFPA 72-1999 3-8.4.1.2

NFPA 72-1999 3-8.4.1.2.1

Distinctive Evacuation Signal. To meet the requirements of NFPA 72-1999 1-5.4.7 (**Distinctive Signals**), the fire alarm signal used to notify building occupants of the need to evacuate (leave the building) will be in accordance with ANSI S3.41, *Audible Emergency Evacuation Signal.*

NFPA 72-1999 3-8.4.1.2.2

The use of the American National Standard Audible Emergency Evacuation Signal will be restricted to situations where it is desired that all occupants hearing the signal evacuate the building immediately. It will not be used where, with the approval of the authority having jurisdiction, the planned action during a fire emergency is not evacuation, but is the relocation of occupants or their protection in place as directed by the firefighting personnel.

NFPA 72-1999 A-3-4.1.2

NFPA 72-1999 Figures A-3-8.4.1.2(a)(b)(c)

The use of the distinctive three-pulse temporal pattern fire alarm evacuation signal required by NFPA 72-1999 3-8.4.1.2.2 became effective July 1, 1996, for new systems installed after that date. It had previously been recommended for this purpose by this code since 1979. It has since been adopted as both an American National Standard (ANSI S3.41, *Audible Emergency Evacuation* Signal) and an International Standard (ISO 8201, *Audible Emergency Evacuation Signal*).

Copies of both of these standards are available from the Standards Secretariat, Acoustical Society of American, 335 East 45th Street, New York, NY 10017-3483. Telephone 212-661-9404 ext. 562.

The standard fire alarm evacuation signal is a three-pulse temporal pattern using any appropriate sound. The pattern consists of an "on" phase (a) lasting 0.5 second ˅ 10 percent followed by and "off" phase (b) lasting 0.5 second ˅ 10 percent, for three successive "on" periods, which are then followed by and "off" phase (c) lasting 1.5 second ˅ 10 percent [*see NFPA 72-1999 Figures A-3-8.4.1.2 (a) and A-3-8.4.1.2 (b)*]. The signal should be repeated for a period appropriate for the purposes of evacuation of the building, but for not less than 180 seconds. A single-stroke bell or chime sounded at "on" intervals lasting 1 second ˅ 10 percent, with a 2-second ˅ 10 percent "off" interval after each third "on" stroke, may be permitted [*see NFPA 72-1999 Figure A-3-8.4.1.2(c)*].

The minimum repetition time is permitted to be manually interrupted.

NFPA 72-1999 3-4.2

Initiating, notification, and signaling line circuits will be designated by class or style, or both, depending upon the circuit's capability to continue to operate during a specified fault condition.

See **Table 3-5 Performance of Initiating Device Circuits (IDS)**

See **Table 3-6 Performance of Signaling Line Circuits (SLC)**

See **Table 3-7.1 Notification Appliance Circuits (NAC)**

NFPA 72-1999 3-8.2.3.1

Systems equipped with alarm verification features are permitted under the following conditions:

1. The alarm verification feature is not initially enabled unless conditions or occupant activities that are expected to cause nuisance alarms are anticipated in the area that is protected by the smoke detectors. Enabling of the alarm verification feature will be protected by password or limited access.

2. A smoke detector continuously subjected to a smoke concentration above alarm threshold does not delay the system functions of 1-5.4 by more than 1 minute.

3. Actuation of an alarm-initiating device other than a smoke detector causes the system functions of 1-5.4 without additional delay.

4. When the alarm verification feature is enabled, disabled, or changed, the comments section of the Record of Completion (Figure 1-6.2.1 item 10) will be used to record the status or change to system operation.

NFPA 72-1999 3-8.3.2.2

Automatic Fire Alarm Signal Initiation. Automatic fire alarm-signal initiating devices having integral trouble contacts will be connected to the initiating device circuit so that a trouble condition within a device does not impair the alarm transmission from any other initiating device, except where the trouble condition is caused by electrical disconnection of the device or by removing the initiating device from its plug-in base.

NFPA 72-1999 3-8.3.2.3.3

Systems that require the operation of two automatic detection devices to initiate the alarm response are permitted as follows:

(a) The systems are not prohibited by the authority having jurisdiction.

> (b) There are at least two automatic detection devices in each protected space.
>
> (c) The alarm verification feature is not used.

NFPA 72-1999 3-8.4.2.2 — **Concealed Detectors.** If a remote alarms indicator is provided for an automatic fire detector in a concealed location, the location of the detector and the area protected by the detector will be prominently indicated at the remote alarm indicator by a permanently attached placard or by other approved means.

NFPA 72-1999 3-8.3.2.3.2 — **Automatic Drift Compensation.** If automatic drift compensation of sensitivity for a fire detector is provided, the control unit will identify the affected detector when the limit of compensation is reached.

NFPA 72-1999 3-8.3.2.4.2 — The number of waterflow switches permitted to be connected to a single initiating device circuit will not exceed five.

NFPA 72-1999 3-8.3.3.1.1 — The number of supervisory devices permitted to be connected to a single initiating device circuit will not exceed 20.

NFPA 72-1999 3-2.2 — **Signal Annunciation.** Protected premises fire alarm systems will be arranged to annunciate alarm, supervisory, and trouble signals in accordance with 1-5.7.

NFPA 72-1999 3-8.2.1 — Fire alarm systems are permitted to share components, equipment, circuitry, and installation wiring with non-fire alarm systems.

NFPA 72-1999 3-8.2.5 — In combination systems, fire alarm signals will be distinctive, clearly recognizable, and take precedence over any other signal even when a non-fire alarm signal in initiated first.

NFPA 72-1999 3-8.5.2 — The number of guard's tour reporting stations, their locations, and the route to be followed by the guard for operating the stations will be approved for the particular installation in accordance with NFPA 601, *Standard for Security Services in Fire Loss Prevention.*

NFPA 72-1999 3-9.3.1

System-type smoke detectors located in elevator lobbies, elevator hoistways, and elevator machine rooms used to initiate fire fighters' service recall will be connected to the building fire alarm system. In facilities without a building fire alarm system, these smoke detectors will be connected to a dedicated fire alarm system control unit that will be designated as "elevator recall control and supervisory panel", permanently identified on the control unit and on the record drawings. Unless otherwise required by the authority having jurisdiction, only the elevator lobby, elevator hoistway, and the elevator machine room smoke detectors are to be used to recall elevators for fire fighters' service.

33017 AUXILIARY FIRE ALARM SYSTEMS

REFERENCES:

NFPA 72

Fire Protection Handbook

Fire Alarm
Signaling Systems

DESCRIPTION:

Understand an auxiliary protective signaling system, including its purpose, types of auxiliary alarm systems, its application, basic requirements, special wiring requirements, power supply sources and types of signaling services available.

NFPA 72-1999 1-4

Auxiliary Fire Alarm System. A system connected to a municipal fire alarm system for transmitting an alarm of fire to the public fire service communications center. Fire alarms from an auxiliary fire alarm system are received at the public fire service communications center on the same equipment and by the same methods as alarms transmitted manually from municipal fire alarm boxes located on streets.

NFPA 72-1999 1-5.2.6

Secondary Supply Capacity and Sources. The secondary supply will automatically supply the energy to the system within 30 seconds, and without loss of signals, wherever the primary supply is incapable of providing the minimum voltage required for proper operation. The secondary (standby) power supply will supply energy to the system in the event of total failure of the primary (main) power supply or when the primary voltage drops to a level insufficient to maintain functionality of the control equipment and system components. **Under maximum normal load, the secondary supply will have sufficient capacity to operate** a protected premises, central station, or proprietary system for 24 hours, or **an auxiliary** or remote station **system for 60 hours; and, at the end of that period, will be capable of operating all alarm notification appliances used for evacuation or to direct aid to the location of an emergency for 5 minutes.** The secondary power supply for emergency voice/alarm communications service will be capable of operating the system under maximum normal load for 24 hours and then be capable of operating the system during a fire, or other emergency condition for a period of 2 hours. Fifteen minutes of evacuation alarm operation at maximum connected load is considered the equivalent of 2 hours of emergency operation.

NFPA 72-1999 6-16.2.2

Permission for the connection of an auxiliary fire alarm system to a public fire alarm reporting system, and acceptance of the type of auxiliary transmitter and its actuating mechanism, circuits, and components connected thereto, is to be obtained from the authority having jurisdiction.

NFPA 72-1999 6-16.2.3 An auxiliary fire alarm system will be maintained and supervised by a responsible person or corporation.

NFPA 72-1999 6-16.2.4 Section 6-16 does not require the use of audible alarm signals other than those necessary to operate the auxiliary fire alarm system. If it is desired to provide fire alarm evacuation signals in the protected property, the alarms, circuits, and controls will comply with the provisions of Chapter 3 in addition to the provisions of Section 6-16.

NFPA 72-1999 6-16.4.1 **Types of Auxiliary Systems.** Auxiliary fire alarm systems are of the following two types:

(a) Local Energy Type.
1. Local energy systems are permitted to be of the coded or non-coded type.
2. Power supply sources for local energy systems will conform to Chapter 1.

(a) Shunt Type.
1. Shunt systems will be non-coded with respect to any remote electrical tripping or actuating devices.
2. All conductors of the shunt circuit will be installed in accordance with NFPA 70, *National Electrical Code*, Article 346, for rigid conduit, or Article 348, for electrical metallic tubing.
3. Both sides of the shunt circuit will be in the same conduit.
4. If an auxiliary transmitter is located within private premises, it will be installed in accordance with 6-9.1.
5. If a shunt loop is used, it will not exceed a length of 750 feet and will be in a conduit.
6. Conductors of the shunt circuits will not be smaller than No. 14 AWG and will be insulated as prescribed in NFPA 70, *National Electrical Code*, Article 310.
7. The power for shunt-type systems will be provided by the public fire alarm reporting system.
8. A local system made to an auxiliary system by the addition of a relay whose coil is energized by a local power supply and who's normally closed contacts trip a shunt-type master box is not permitted.

NFPA 72-1999 6-16.4.4.1 Shunt-type auxiliary systems are to be arranged so that one auxiliary transmitter does not serve more than 100,000 feet^2 total area, except where otherwise permitted by the authority having jurisdiction.

33018 SUPERVISING STATION FIRE ALARM SYSTEMS

REFERENCES:

NFPA 72

NFPA 101

Fire Alarm
Signaling Systems

DESCRIPTION:

Understand a remote station protective signaling system, including its purpose, system operation, general requirements, power supplies at protected premises and the remote station, signal transmission, operation under fault condition and types of services available.

NFPA 72-1999 1-4

Remote Supervising Station Fire Alarm Systems. A system installed in accordance with NFPA 72-1999 to transmit alarm, supervisory, and trouble signals from one or more protected premises to a remote location where appropriate action is taken.

NFPA 72-1999 1-5.2.6

Secondary Supply Capacity and Sources. The secondary supply will automatically supply the energy to the system within 30 seconds, and without loss of signals, wherever the primary supply is incapable of providing the minimum voltage required for proper operation. The secondary (standby) power supply will supply energy to the system in the event of total failure of the primary (main) power supply or when the primary voltage drops to a level insufficient to maintain functionality of the control equipment and system components. **Under maximum normal load, the secondary supply will have sufficient capacity to operate a protected premises, central station, or proprietary system for 24 hours, or an auxiliary or remote station system for 60 hours, and, at the end of that period, will be capable of operating all alarm notification appliances used for evacuation or to direct aid to the location of an emergency for 5 minutes.** The secondary power supply for emergency voice/alarm communications service will be capable of operating the system under maximum normal load for 24 hours and then be capable of operating the system during a fire, or other emergency condition for a period of 2 hours. Fifteen minutes of evacuation alarm operation at maximum connected load is considered the equivalent of 2 hours of emergency operation.

NFPA 72-1999 5-4

NFPA 72-1999 5-4.1

Remote Supervising Station Fire Alarm Systems.
Scope. NFPA 72-1999 5-4 applies where central station service is neither required nor elected. Section 5-4 describes the installation, maintenance, testing, and use of a remote supervising station fire alarm system that serves properties under various ownership from a

remote supervising station where trained, competent personnel are in constant attendance. Section 5-4 covers the minimum requirements for the remote supervising station physical facilities, equipment, operating personnel, response, retransmission, signals, reports, and testing.

NFPA 72-1999 A-5-4.3

As a minimum, the room or rooms containing the remote supervising station equipment should have a 1-hour fire rating, and the entire structure should be protected by a fire alarm system complying with NFPA 72-1999 Chapter 3.

NFPA 72-1999 5-4.3.1

If a remote supervising station connection is used to transmit an alarm signal, the signal will be received at the public fire service communications center, at a fire station, or at the governmental agency that has a public responsibility for taking prescribed action to ensure response upon receipt of a fire alarm signal.

If such an agency is unwilling to receive alarm signals or permits the acceptance of another location by the authority having jurisdiction, such alternate location will have personnel on duty at all times who are trained to receive the alarm signal and immediately retransmit it to the fire department.

NFPA 72-1999 5-4.3.2

Supervisory and trouble signals are to be handled at a constantly attended location that has personnel on duty who are trained to recognize the type of signal received and to take prescribed action. The location is permitted to be other than that at which alarm signals are received.

NFPA 72-1999 5-4.3.3

If locations other than the public fire service communications center are used for the receipt of signals, access to receiving equipment will be restricted in accordance with requirements of the authority having jurisdiction.

NFPA 72-1999 5-4.4.1

Signal-receiving equipment will indicate receipt of each signal both audibly and visibly.

NFPA 72-1999 5-4.4.1.2

Means for silencing alarm, supervisory, and trouble signals are to be provided and will be arranged so that subsequent signals will re-sound.

NFPA 72-1999 5-4.4.1.3 A trouble signal will be received when the system or any portion of the system at the protected premises is placed in a bypass or test mode.

NFPA 72-1999 5-4.4.1.4 An audible and visible indication will be provided upon restoration of the systems after receipt of any signal.

NFPA 72-1999 5-4.4.2 Power supplies will comply with the requirements of NFPA 72-1999 Chapter 1.

rad.b In a remote supervising station fire alarm system where the alarm and supervisory signals are transmitted over a listed supervised one-way cable system, 24 hours of secondary (standby) power is permitted in lieu of 60 hours, as required in 1-5.2.5.3, at the radio alarm repeater station receivers (RARSR), provided that personnel are dispatched to arrive within 4 hours after detection of failure to initiate maintenance.

NFPA 72-1999 5-4.4.4 Retransmission of an alarm signal, if required, will be by one of the following methods, which appear in descending order of preference as follows:

(a) A dedicated circuit that is independent of any switched telephone network. This circuit is permitted to be used for voice or data communications.

(b) A one-way (outgoing only) telephone at the remote supervising station that utilizes the public switched telephone network. This telephone is to be used primarily for voice transmission of alarms to a telephone at the public fire service communications center that cannot be used for outgoing calls.

(c) A private radio system using the fire department frequency, where permitted by the fire department.

(d) Other methods acceptable to the authority having jurisdiction.

NFPA 72-1999 5-4.6.1 If the remote supervising station is at a location other than the public fire service communications center, alarm signals will be immediately retransmitted to the public fire service communications center.

NFPA 72-1999 5-4.6.2 Upon receipt of an alarm, supervisory, or trouble signal by the remote supervising station other than the public fire service communications center, the operator on duty is responsible for notifying the owner or the owner's designated representative immediately.

NFPA 72-1999 5-4.7.1   A permanent record of the time, date, and location of all signals and restorations received; the action taken and the actions taken will be maintained for at least 1 year and made available to the authority having jurisdiction. These records are permitted to be created by manual means.

33019 PROPRIETARY SUPERVISING STATION SYSTEMS

REFERENCES:

NFPA 72

DESCRIPTION:

Understand a proprietary fire alarm system, including its purpose, system operation, general requirements, power supplies at central proprietary station and remotely located control equipment, signal transmission and processing, types of service available, the performance capacities of initialing device and signaling line circuits and alarm retransmissions.

NFPA 72-1999 1-4

Proprietary Supervising Station Fire Alarm System. An installation of fire alarm systems that serves contiguous and noncontiguous properties, under one ownership, from a proprietary supervising station located at the protected property, at which trained, competent personnel are in constant attendance. This includes the proprietary supervising station; power supplies; signal-initiating devices; initiating device circuits; signal notification appliances; equipment for the automatic, permanent visual recording of signals; and equipment for initiating the operation of emergency building control services.

NFPA 72-1999 1-5.2.6

Secondary Supply Capacity and Sources. The secondary supply will automatically supply the energy to the system within 30 seconds, and without loss of signals, wherever the primary supply is incapable of providing the minimum voltage required for proper operation. The secondary (standby) power supply will supply energy to the system in the event of total failure of the primary (main) power supply or when the primary voltage drops to a level insufficient to maintain functionality of the control equipment and system components. **Under maximum normal load, the secondary supply will have sufficient capacity to operate a** protected premises, central station, or **proprietary system for 24 hours**, or an auxiliary or remote station system for 60 hours; **and, at the end of that period, will be capable of operating all alarm notification appliances used for evacuation or to direct aid to the location of an emergency for 5 minutes.** The secondary power supply for emergency voice/alarm communications service will be capable of operating the system under maximum normal load for 24 hours and then be capable of operating the system during a fire, or other emergency condition for a period of 2 hours. Fifteen minutes of evacuation alarm operation at maximum connected load is considered the equivalent of 2 hours of emergency operation.

NFPA 72-1999 5-3.3.1

The proprietary supervising station is to be located in a fire-resistive, detached building or in a cutoff room and will not be exposed to the hazardous parts of the premises that are protected.

NFPA 72-1999 5-3.3.2

Access to the proprietary supervising station will be restricted to those persons directly concerned with the implementation and direction of emergency action and procedure.

NFPA 72-1999 5-3.3.3

The proprietary supervising station, as well as remotely located power rooms for batteries or engine-driven generators, will be provided with portable fire extinguishers that comply with requirements of NFPA 10, *Standard for Portable Fire Extinguishers*.

NFPA 72-1999 5-3.3.4.1

The proprietary supervising station is to be provided with an automatic emergency lighting system. The emergency source will be independent of the primary lighting source.

NFPA 72-1999 5-3.3.4.2

In the event of a loss of primary lighting for the supervising station, the emergency lighting will provide illumination for a period of not less than 26 hours to permit the operators to carry on operations, and will be tested in accordance with the requirements of NFPA 72-1999 Chapter 7.

NFPA 72-1999 5-3.3.5

If 25 or more protected buildings or premises are connected to a subsidiary station, both of the following will be provided at the subsidiary station:

    (a) Automatic means for receiving and recording signals under emergency-staffing conditions; and
    (b) A telephone.

NFPA 72-1999 5-3.4.2

Provision is to be made to designate the building in which a signal originates. The floor, section, or other subdivision of the building will be designated at the proprietary supervising station or at the building that is protected except where the area, height, or special conditions of occupancy make detailed designation unessential as approved by the authority having jurisdiction. This detailed designation will use indicating appliances accepted by the authority having jurisdiction.

NFPA 72-1999 5-3.4.7

The maximum elapsed time from sensing a fire alarm at an initiating device or initiating device circuit until it is recorded or displayed at the proprietary supervising station will not exceed 90 seconds.

NFPA 72-1999 5-3.4.8

To facilitate the prompt receipt of fire alarm signals from systems handling other types of signals that are able to produce multiple simultaneous status changes, the requirements of either of the following will be met:

(a) In addition to the maximum processing time for a single alarm, the system will record simultaneous status changes at a rate not slower than either a quantity of 50, or 10 percent of the total number of initiating device circuits connected, within 90 seconds, whichever number is smaller, without loss of any signal; or,

(b) In addition to the maximum processing time, the system will either display or record fire alarm signals at a rate not slower than one every 10 seconds, regardless of the rate or number of status changes occurring, without loss of any signals.

If fire alarm, waterflow alarm, and sprinkler supervisory signals and their associated trouble signals are the only signals processed by the system, the rate of recording will not be slower than one signal every 30 seconds.

NFPA 72-1999 5-3.4.9

Trouble signals required by 1-5.8 and their restoration will be automatically indicated and recorded at the proprietary supervising station within 200 seconds.

NFPA 72-1999 5-3.5.1

At least two operators will be on duty at all times. One of the two operators is permitted to be a runner.

If the means for transmitting alarms to the fire department is automatic, at least one operator will be on duty at all times.

NFPA 72-1999 5-3.5.2

When the runner is not in attendance at the proprietary supervising station, the runner will establish two-way communications with the station at intervals not exceeding 15 minutes.

NFPA 72-1999 5-3.5.3

The primary duties of the operator(s) is to monitor signals, operate the system, and take such action as is required by the authority having jurisdiction. The operator(s) will not be assigned any additional duties that would take precedence over the primary duties.

NFPA 72-1999 5-3.6.6.1

Alarms. Upon receipt of a fire alarm signal, the proprietary supervising station operator will initiate action to perform the following:

    (a) Immediately notify the fire department, the plant fire brigade, and such other parties as the authority having jurisdiction requires.

    (b) Promptly dispatch a runner to the alarm location (travel time will not exceed 1 hour).

    (c) Restore the system as soon as possible after disposition of the cause of the alarm signal.

NFPA 72-1999 5-3.6.7.1

Complete records of all signals received will be retained for at least 1 year.

33020 CENTRAL STATION FIRE ALARM SYSTEMS

REFERENCES:

NFPA 72

DESCRIPTION:

Understand a central station signaling system, including its purpose, operation, basic requirements, power supply sources at the central station and protected premises, signal capacity of circuits, types of signaling services available, active multiplex systems, digital alarm communicator systems and alarm retransmission.

NFPA 72-1999 1-4

Central Station Fire Alarm System. A system or group of systems in which the operations of circuits and devices are transmitted automatically to, recorded in, maintained by, and supervised from a listed central station that has competent and experienced servers and operators who, upon receipt of a signal, take such action as required by NFPA 72-1999. Such service is to be controlled and operated by a person, firm, or corporation whose business is the furnishing, maintaining, or monitoring of supervised fire alarm systems.

NFPA 72-1999 1-5.2.6

Secondary Supply Capacity and Sources. The secondary supply will automatically supply the energy to the system within 30 seconds, and without loss of signals, wherever the primary supply is incapable of providing the minimum voltage required for proper operation. The secondary (standby) power supply will supply energy to the system in the event of total failure of the primary (main) power supply or when the primary voltage drops to a level insufficient to maintain functionality of the control equipment and system components. **Under maximum normal load, the secondary supply will have sufficient capacity to operate** a protected premises, **central station**, or proprietary system for 24 hours, or an auxiliary or remote station **system for 60 hours; and, at the end of that period, will be capable of operating all alarm notification appliances used for evacuation or to direct aid to the location of an emergency for 5 minutes.** The secondary power supply for emergency voice/alarm communications service will be capable of operating the system under maximum normal load for 24 hours and then be capable of operating the system during a fire, or other emergency condition for a period of 2 hours. Fifteen minutes of evacuation alarm operation at maximum connected load is considered the equivalent of 2 hours of emergency operation.

NFPA 72-1999 5-2

Fire Alarm Systems for Central Station Service.

NFPA 72-1999 5-2.2.2

NFPA 72-1999 Section 5-2 applies to central station service, which consists of the following elements:

- (1) Installation of fire alarm transmitters
- (2) Alarm, guard, supervisory, and trouble signal monitoring
- (3) Retransmission
- (4) Associated record keeping and reporting
- (5) Testing and maintenance
- (6) Runner service

The central station service elements will be provided under contract to a subscriber by one of the following:

(a) A listed central station that provides all of the elements of central station service with its own facilities and personnel; or

(b) A listed central station that provides, as a minimum, the signal monitoring, retransmission, and associated record keeping and reporting with its own facilities and personnel and is permitted to subcontract all or any part of the installation, testing and maintenance, and runner service; or

(c) A listed fire alarm service-local company that provides the installation and testing and maintenance with its own facilities and personnel and that subcontracts the monitoring, retransmission, and associated record keeping and reporting to a listed central station. The required runner service is to be provided by the listed fire alarm service-local company with its own personnel or the listed central station with its own personnel.

NFPA 72-1999 5-2.2.3.1

The installation will be certificated.

NFPA 72-1999 5-2.2.3.2

The installation will be placarded.

NFPA 72-1999 5-2.2.3.2.2 — The placard(s) will be 20 inches^2 or larger, located on or within 36 inches of the fire alarm system control unit or, if no control unit exists, on or within 36 inches of a fire alarm system component, and will identify the central station by name and telephone number.

NFPA 72-1999 5-2.3.1 — The central station building or that portion of a building occupied by a central station will conform to the construction, fire protection, restricted access, emergency lighting, and power facilities, requirements of the latest edition of ANSI/UL 827, *Standard for Safety Central-Station for Watchman, Fire-Alarm and Supervisory Services.*

NFPA 72-1999 5-2.3.2 — Subsidiary station buildings or those portions of buildings occupied by subsidiary stations will conform to the construction, fire protection, restricted access, emergency lighting, and power facilities requirements of the latest edition of ANSI/UL 827, *Standard for Safety Central-Station for Watchman, Fire-Alarm and Supervisory Services.*

NFPA 72-1999 5-2.3.2.1 — All intrusion, fire, power, and environmental control systems for subsidiary station buildings will be monitored by the central station in accordance with 5-2.3.

NFPA 72-1999 5-2.3.2.2 — The subsidiary facility will be inspected at least monthly by central station personnel for the purpose of verifying the operation of all supervised equipment, all telephones, all battery conditions, and all fluid levels of batteries and generators.

NFPA 72-1999 5-2.3.2.3 — In the event of the failure of equipment at the subsidiary station or the communications channel to the central station, a backup will be operational within 90 seconds. Restoration of a failed unit will be accomplished within 5 days.

NFPA 72-1999 5-2.4.4 — Two independent means are to be provided to retransmit a fire alarm signal to the designated public fire service communications center.

NFPA 72-1999 5-2.5.1

The central station will have sufficient personnel, but not less than two persons, on duty at the central station at all times to ensure disposition of signals in accordance with the requirements of NFPA 72-1999 5-2.6.1.

NFPA 72-1999 5-2.5.2

Operation and supervision will be the primary functions of the operators, and no other interest or activity will take precedence over the protective service.

NFPA 72-1999 5-2.6.1.1

Alarm signals initiated by manual fire alarm boxes, automatic fire detectors, waterflow from the automatic sprinkler system, or actuation of other fire suppression system(s) or equipment will be treated as fire alarms.
The central station will perform all of the following actions:

(1) Immediately retransmit the alarm to the public fire service communications center

(2) Dispatch a runner or technician to the protected premises to arrive within 1 hour after receipt of a signal if equipment needs to be manually reset by the prime contractor

(3) Immediately notify the subscriber

(4) Provide notice to the subscriber or authority having jurisdiction, or both, if required.

If the alarm signal results from a prearranged test, the actions specified by 5-2.6.1.1(1) and (3) are not required.

NFPA 72-1999 5-2.6.2.1

Complete records of all signals received will be retained for at least 1 year.

NFPA 72-1999 5-2.6.2.3

The central station will make arrangements to furnish reports of signals received to the authority having jurisdiction in a manner approved by the authority having jurisdiction.

33021 MANUAL FIRE ALARM SYSTEMS & GUARD'S TOUR SERVICE

REFERENCES:

NFPA 72

Basic/National Building Code

Standard Building Code

International Building Code

DESCRIPTION:

Understand a manual fire alarm system, including location, mounting and distribution of manual fire alarm boxes, use and operation of combination manual fire alarm and guard's tour stations. Understand special requirements for guard's tour supervisory service. Lay out a combination manual fire alarm and guard's tour system based on code requirements. Know the operating principles of noncoded, coded, breakglass, nonbreakglass, pre-signal, general alarm, single action and double action fire alarm stations and in what type of system each would be used.

NFPA 72-1999 1-4

Combination Fire Alarm and Guard's Tour Box. A manually operated box for separately transmitting a fire alarm signal and a distinctive guard patrol tour supervisory signal.

NFPA 72-1999 1-4

Guard's Tour Supervisory Signal. A supervisory signal monitoring the performance of guard patrols.

NFPA 72-1999 1-4

Guard's Tour Reporting Station. A device that is manually or automatically initiated to indicate the route being followed and the timing of a guard's tour.

NFPA 72-1999 1-5.4.10

Presignal Feature. Where permitted by the authority having jurisdiction, systems are permitted to have a feature that allows initial fire alarm signals to sound only in department offices, control rooms, fire brigade stations, or other constantly attended central locations and for which human action is subsequently required to activate a general alarm, or a feature that allows the control equipment to delay the general alarm by more than 1 minute after the start of the alarm processing. If there is a connection to a remote location, the transmission of the alarm signal to the supervising station will activate upon the initial alarm signal.

NFPA 72-1999 A-1-5.4.10

A system provided with an alarm verification feature as permitted by 3-8.3.2.3.1 is not considered a presignal system, since the delay in the signal produced is 60 seconds or less and requires no human intervention.

NFPA 72-1999 5-2.6.1.2.1

Upon failure to receive a guard's tour supervisory signal within a 15-minute maximum grace period, the central station will perform the following actions:

    (a) Communicate without unreasonable delay with personnel at the protected premises

    (b) Dispatch a runner to the protected premises to arrive within 30 minutes of the delinquency if communications cannot be established

NFPA 72-1999 3-8.5
HANDBOOK

Guard's Tour Supervisory Service.

Guard's tour supervisory service is used to provide fire protection surveillance during the hours when occupants are not in a building; to facilitate and control the movement of persons into, out of, and within a building; and to carry out procedures for the orderly conduct of specific operations in a building or on the surrounding property.

Guard's tour supervisory services designed to continually report the performance of a guard are often found in connection with protected premises fire alarm systems utilizing off-premises reporting through central or proprietary supervising stations.

NFPA 72-1999 3-8.5.2

The number of guard's tour reporting stations, their locations, and the route to be followed by the guard for operating the stations are to be approved for the particular installation in accordance with NFPA 601, *Standard for Security Services in Fire Loss Prevention.*

NFPA 72-1999 3-8.5.3

A permanent record indicating every time each signal-transmitting station is operated will be made at the main control unit. Where intermediate stations that do not transmit a signal are employed in conjunction with signal-transmitting stations, distinctive signals will be transmitted at the beginning and end of each tour of a guard, and a signal-transmitting station will be provided at intervals not exceeding 10 stations. Intermediate stations that do not transmit a signal will be capable of operation only in a fixed sequence.

NFPA 72-1999 2-8 — Manual fire alarm boxes will be used only for fire alarm-initiating purposes. However, combination manual fire alarm boxes and guard's signaling stations are permitted.

NFPA 72-1999 2-8.1

 Each manual fire alarm box will be securely mounted. The operable part of each manual pull box will not be less than 3 1/2 feet and not more than 4 1/2 feet above floor level.

NFPA 72-1999 2-8.2.1 — Manual fire alarm boxes will be distributed throughout the protected area so that they are unobstructed and accessible.

NFPA 72-1999 2-8.2.2 — Manual fire alarm boxes will be located within 5 feet of the exit doorway opening at each exit on each floor.

NFPA 72-1999 2-8.2.3 — Manual fire alarm boxes will be mounted on both sides of group openings over 40 feet in width. Manual fire alarm boxes will be mounted within 5 feet of each side of the opening.

NFPA 72-1999 2-8.2.4 — Additional manual fire alarm boxes will be provided so that travel distance to the nearest fire alarm box will not be in excess of 200 feet measured horizontally on the same floor.

NFPA 72-1999 2-8.3 — A coded manual fire alarm box will produce at least three repetitions of the coded signal, with each repetition to consist of at least three impulses.

Notes

33022 HEAT SENSING FIRE DETECTORS

REFERENCES:

NFPA 72

*Fire Alarm
Signaling Systems*

*UL Fire Protection
Equipment Directory*

DESCRIPTION:

Understand the basic principles of operation of fixed temperature, rate-of-rise, rate compensation and combination heat sensing fire detectors, their temperature classifications, listed spacing and the effect of thermal lag. Lay out a system of heat sensing fire detectors of both the line and spot type.

NFPA 72-1999 1-4

Heat Detector. A fire detector that detects either abnormally high temperature or rate of temperature rise, or both.

NFPA 72-1999 1-4

NFPA 72-1999A-1-4

Fixed Temperature Detector. A device that responds when its operating element becomes heated to a predetermined level.

The difference between the operating temperature of a fixed temperature device and the surrounding air temperature is proportional to the rate at which the temperature is rising. The rate is commonly referred to as *thermal lag*. The air temperature is always higher than the operating temperature of the device.

Typical examples of fixed temperature-sensing elements are as follows:

(a) *Bimetallic*. A sensing element comprised of two metals having different coefficients of thermal expansion arranged so that the effect is deflection in one direction when heated and in the opposite direction when cooled;

(b) *Electrical Conductivity*. A line-type or spot-type sensing element whose resistance varies as a function of temperature;

(c) *Fusible Alloy*. A sensing element of a special composition (eutectic) metal that melts rapidly at the rated temperature;

(d) *Heat-Sensitive Cable*. A line-type device whose sensing element comprises, in one type, two current-carrying wires separated by heat-sensitive insulation that softens at the rated temperature, thus allowing the wires to make electrical contact. In another type, a single wire is centered in a metallic tube, and the intervening space is filled with a substance that, at a critical temperature, becomes conductive, thus establishing electrical contact between the tube and the wire; and

(e) Liquid Expansion. A sensing element comprising a liquid capable of marked expansion in volume in response to temperature increase.

NFPA 72-1999 1-4

NFPA 72-1999 A-1-4

Rate Compensation Detector. A device that responds when the temperature of the air surrounding the device reaches a predetermined level, regardless of the rate of temperature rise.

A typical example is a spot-type detector with a tubular casing of a metal that tends to expand lengthwise as it is heated and an associated contact mechanism that closes at a certain point in the elongation. A second metallic element inside the tube exerts an opposing force on the contacts, tending to hold them open. The forces are balanced in such a way that, on a slow rate-of-temperature rise, there is more time for heat to penetrate to the inner element, which inhibits contact closure until the total device has been heated to its rated temperature level. However, on a fast rate-of-temperature rise, there is not as much time for heat to penetrate to the inner element, which exerts less of an inhibiting effect so that contact closure is achieved when the total device has been heated to a lower temperature. This, in effect, compensates for thermal lag.

NFPA 72-1999 1-4

NFPA 72-1999A-1-4

Rate-of-Rise Detector. A device that responds when the temperature rises at a rate exceeding a predetermined value.

Typical examples of rate-of-rise detectors follow.

(a) Pneumatic Rate-of Rise Tubing. A line-type detector comprising small-diameter tubing, usually copper, that is installed on the ceiling or high on the walls throughout the protected area. The tubing is terminated in a detector unit containing diaphragms and associated contacts set to actuate at a predetermined pressure. The system is sealed except for calibrated vents that compensate for normal changes in temperature.

(b) Spot-Type Pneumatic Rate-of-Rise Detector. A device consisting of an air chamber, a diaphragm, contacts, and a compensating vent in a single enclosure. The principle of operation is the same as that described for pneumatic rate-of-rise tubing.

(c) Electrical Conductivity-Type Rate-of-Rise Detector. A line-type or spot-type sensing element whose resistance changes due to a change in temperature. The rate of change

of resistance is monitored by associated control equipment,and an alarm is initiated when the rate of temperature increase exceeds a preset value.

NFPA 72-1999 2-2.1.1

Color Coding.

NFPA 72-1999 2-2.1.1.1

Heat-sensing fire detectors of the fixed-temperature or rate-compensated spot-type will be classified as to the temperature of operation and marked with the appropriate color code in accordance with NFPA 72-1999 Table 2-2.1.1.1.

NFPA 72-1999 2-2.1.1.2

If the overall color of a heat-sensing fire detector is the same as the color code marking required for that detector, one of the following arrangements, applied in a contrasting color and visible after installation, will be employed:

(a) Ring on the surface of the detector;

(b) Temperature rating in numerals at least 3/8 inch high.

NFPA 72-1999 2-2.1.2

A heat-sensing fire detector integrally mounted on a smoke detector will be listed or approved for not less than 50-feet spacing.

NFPA 72-1999 A-2.2.1.2

The linear space rating is the maximum allowable distance between heat detectors. The linear space rating is also a measure of the heat detector response time to a standard test fire where tested at the same distance. The higher the rating, the faster the response time. NFPA 72-1999 recognizes only those heat detectors with ratings of 50 feet or more.

NFPA 72-1999 2-2.2.1

NFPA 72-1999 Figure A-2-2.2.1

Spot-type heat-sensing fire detectors will be located on the ceiling not less than 4 inches from the sidewall or on the sidewalls between 4 inches and 12 inches from the ceiling.

In the case of solid open joist construction, detectors will be mounted at the bottom of the joists.

In the case of beam construction where beams are less than 12 inches in depth and less than 8 feet on center, detectors are permitted to be installed on the bottom of beams.

NFPA 72-1999 2-2.2.2

Line-type heat detectors will be located on the ceiling or on the sidewalls not more that 20 inches from the ceiling.

In the case of solid open joist construction, detectors will be mounted at the bottom of the joist.

In the case of beam construction where beams are less than 12 inches in depth and less than 8 feet on center, detectors are permitted to be installed on the bottom of beams.

If a line-type detector is used in an application other than open area protection, the manufacturer's installation instructions will be followed.

NFPA 72-1999 2-2.3

Temperature. Heat-sensing fire detectors having fixed-temperature or rate-compensated elements will be selected in accordance with NFPA 72-1999 Table 2-2.1.1.1 for the maximum expected ceiling temperature that can be expected.
The temperature rating of the detector will be at least 20°F above the maximum expected temperature at the ceiling.

NFPA 72-1999 2-2.4.1

Smooth Ceiling Spacing. One of the following requirements apply:

NFPA 72-1999 A-2.2.4.1

(1) The distance between detectors will not exceed their listed spacing, and there will be detectors within a distance of 1/2 the listed spacing, measured at a right angle, from all walls or partitions extending to within 18 inches of the ceiling; or

(2) All points on the ceiling will have a detector within a distance equal to 0.7 times the listed spacing (0.7S). This is useful in calculating locations in corridors or irregular areas; or

In laying out detector installations, designers work in terms of rectangles, as building areas are generally rectangular in shape. The pattern of heat spread from a fire source, however, is not rectangular in shape. On a smooth ceiling, heat spreads out in all directions in an ever-expanding circle. Thus, the coverage of a detector is not, in fact, a square, but rather a circle whose radius is the linear spacing multiplied by 0.7.

NFPA 72-1999 2-2.4.5

High Ceilings.

NFPA 72-1999 2-2.4.5.1

On ceilings 10 feet to 30 feet high, heat detector linear spacing will be reduced in accordance with NFPA 72-1999 Table 2-2.4.5.1, prior to any additional recuctions for beams, joists, or slope, where applicable.

NFPA 72-1999 Table 2-2.4.5.1does not apply to the following detectors which rely on the integration effect:

(a) Line-type electrical conductivity detectors (See A-1-4)

(b) Pneumatic rate-of-rise tubing (See A-1-4)

In these cases, the manufacturer's recommendations are to be followed for appropriate alarm point and spacing.

33023 SMOKE SENSING FIRE DETECTORS

REFERENCES:

NFPA 72, 80, 90A

*Fire Alarm
Signaling Systems*

*UL Fire Protection
Equipment Directory*

*NEMA Guide for Proper
Use of System Smoke
Detectors*

*NEMA Guide for Proper
Use of Smoke Detectors
in Duct Applications*

DESCRIPTION:

Understand the basic principles of operation of ionization, photoelectric obscuration, photoelectric scattering, and cloud chamber smoke sensing fire detectors and the effect of stratification on response time. Lay out a system of smoke sensing fire detectors of line type, spot type, duct type and sampling type including smoke detectors for use as area smoke detectors, air duct smoke detectors and smoke detectors for door release service.

NFPA 72-1999 1-4

Cloud Chamber Smoke Detection. The principle of using an air sample drawn from the protected area into a high-humidity chamber combined with a lowering of chamber pressure to create an environment in which the resultant moisture in the air condenses on any smoke particles present, forming a cloud, The cloud density is measured by a photoelectric principle. The density signal is processed and used to convey an alarm condition when it meets preset criteria.

NFPA 72-1999 1-4

NFPA 72-1999A-1-4

Ionization Smoke Detection. The principle of using a small amount of radioactive material to ionize the air between two differentially charged electrodes to sense the presence of smoke particles. Smoke particles entering the ionization volume decrease the conductance of the air by reducing ion mobility. The reduced conductance signal is processed and used to convey an alarm condition when it meets preset criteria.

Ionization smoke detection is more responsive to invisible particles (smaller than 1 micron in size) produced by most flaming fires. It is somewhat less responsive to the larger particles typical of most smoldering fire. Smoke detectors utilizing the ionization principle are usually of the spot type.

NFPA 72-1999 1-4

NFPA 72-1999A-1-4

Photoelectric Light Obscuration Smoke Detection. The principle of utilizing a light source and a photosensitive sensor onto which the principal portion of the source emissions is focused. When smoke particles enter the light path, some of the light is scattered and some is absorbed, thereby reducing the light reaching the receiving sensor. The light reduction signal is processed and used to convey an alarm condition when it meets preset criteria.

The response of photoelectric light obscuration smoke detectors is usually not affected by the color of smoke. Smoke detectors utilizing the light obscuration principle are usually of the line type. These detectors are commonly refered to as "projected beam smoke detectors."

NFPA 72-1999 1-4

NFPA 72-1999 A-1-4

Photoelectric Light-Scattering Smoke Detection. The principle of utilizing a light source and a photosensitive sensor arranged in a manner so that the rays from the light source do not normally fall onto the photosensitive sensor. When smoke particles enter the light path, some of the light is scattered by reflection and refraction onto the sensor. The light signal is processed and used to convey an alarm condition when it meets preset criteria.

Photoelectric light-scattering smoke detection is more responsive to visible particles (larger than 1 micron in size) produced by most smoldering fires. It is somewhat less responsive to the smaller particles typical of most flaming fires.

NFPA 72-1999 A-1-4

Smoke Detector. A device that detects visible or invisible particles of combustion.

NFPA 72-1999 2-3.3.1

Smoke detectors will be marked with their nominal production sensitivity (percent per foot obscuration), as required by the listing. The production tolerance around the nominal sensitivity also will be indicated.

NFPA 72-1999 2-3.3.2

Smoke detectors that have provision for field adjustment of sensitivity will have an adjustment range of not less than 0.6 percent per foot obscuration. If the means of adjustment is on the detector, a method will be provided to restore the detector to its factory calibration. Detectors that have provision for program- controlled adjustment of sensitivity are permitted to be marked with their programmable sensitivity range only.

NFPA 72-1999 2-3.4.1.1

The location and spacing of smoke detectors will result from an evaluation based on the guidelines detailed in this code and on engineering judgement. Some of the conditions that will be included in the evaluation include:

    a.) Ceiling shape and surface
    b.) Ceiling height
    c.) Configuration of contents in the area to be protected
    d.) Burning characteristics of the combustible materials present
    e.) Ventilation
    f.) Ambient environment

NFPA 72-1999 2-3.4.1.2

If the intent is to protect against a specific hazard, the detector(s) are permitted to be installed closer to the hazard in a position where the detector can intercept the smoke.

NFPA 72-1999 2-3.4.2

Air Sampling-type Smoke Detector. Each sampling port of an air sampling-type smoke detector will be treated as a spot-type detector for the purpose of location and spacing. Maximum air sample transport time from the farthest sampling point will not exceed 120 seconds.

NFPA 72-1999 2-3.4.3.1

NFPA 72-1999 Figure A-2.2.2.1

Spot-type smoke detectors will be located on the ceiling not less than 4 inches from a sidewall to the near edge or, where on a sidewall, between 4 inches and 12 inches down from the ceiling to the top of the detector.

NFPA 72-1999 2-3.4.3.2

NFPA 72-1999 Figure A-2-3.4.3.2

To minimize dust contamination, smoke detectors, where installed under raised floors, will be mounted only in an orientation for which they have been listed.

NFPA 72-1999 2-3.4.4

Projected Beam-Type Smoke Detectors. Projected beam-type smoke detectors will be located with their projected beams parallel to the ceiling and in accordance with the manufacturer's documented instructions. The effects of stratification will be evaluated when locating the detectors.

Beams are permitted to be installed vertically or at any angle needed to afford protection of the hazard involved (for example, vertical beams through the open shaft area of a stairwell where there is a clear vertical space inside the handrails).

NFPA 72-1999 2-3.4.4.1

The beam length will not exceed the maximum permitted by the equipment listing.

NFPA 72-1999 2-3.4.5.1.1   **Smooth Ceiling Spacing of Spot-Type Detectors.**

On smooth ceilings, spacing of 30 feet is permitted to be used as a guide. In all cases, the manufacturer's documented instructions will be followed. Other spacing is permitted to be used depending on ceiling height, different conditions, or response requirements. For the detection of flaming fires, the guidelines in NFPA 72-1999 Appendix B is permitted to be used.

NFPA 72-1999 2-3.4.5.1.2   For smooth ceilings, all points on the ceiling will have a detector within a distance equal to 0.7 times the selected spacing.

NFPA 72-1999 2-3.4.6.1

Flat Ceilings.
For ceiling heights of 12 feet or lower and beam or solid joists depths of 1 foot or less, smooth ceiling spacing running in the direction parallel to the run of the beams or solid joists will be used and 1/2 the smooth ceiling spacing will be in the direction perpendicular to the run of the beams or solid joists. For beam over 1 foot in depth, spot-type detectors are permitted to be located either on the ceiling or on the bottom of the beams.

For beam depths exceeding 1 foot or for ceiling heights exceeding 12 feet, spot-type detectors will be located on the ceiling in every beam pocket.

For solid joists, the detectors will be located on the bottom of the joists.

NFPA 72-1999 2-3.4.7

**NFPA 72-1999
Figure A-2-2.4.4.1**

Peaked. Detectors will first be spaced and located within 3 feet of the peak, measured horizontally. The number and spacing of additional detectors, if any, will be based on the horizontal projection of the ceiling.

NFPA 72-1999 2-3.4.8

**NFPA 72-1999
Figure A-2-2.4.4.2**

Shed. Detectors will first be spaced and located within 3 feet of the high side of the ceiling, measured horizontally. The number and spacing of additional detectors, if any, will be based on the horizontal projection of the ceiling.

NFPA 72-1999 2-3.4.10

Partitions. Where partitions extend upward to within 18 inches of the ceiling, they will not influence the spacing. Where the partition extends to within less than 18 inches of the ceiling, the effect of smoke travel will be evaluated in the reduction of spacing.

NFPA 72-1999 A-2.3.5.1

Detectors should not be located in a direct airflow nor closer than 3 feet from an air supply diffuser or return air opening. Supply or return sources larger than those commonly found in residential and small commercial establishments can require greater clearance to smoke detectors. Similarly, smoke detectors should be located farther away from high velocity air supplies.

NFPA 72-1999 2-3.5.2.2

Detectors placed in environmental air ducts or plenums will not be used as a substitute for open area detectors. If detectors are used for the control of smoke spread, the requirements of NFPA 72-1999 Section 2-10 will apply.

NFPA 72-1999 2-3.6.1.1

Smoke detectors are not to be installed if any of the following ambient conditions exist:

(a) Temperature below 32°F

(b) Temperature above 100°F

(c) Relative humidity above 93 percent

(d) Air velocity greater than 300 feet per minute

Except detectors specifically designed for use in ambient conditions exceeding the limits of 2-3.6.1.1(a) through (d) and listed for the temperature, humidity, and air velocity conditions expected.

NFPA 72-1999 2-3.6.1.3

Detectors will not be installed until after the construction clean-up of all trades is complete and final, except where required by the authority having jurisdiction for protection during construction. Detectors that have been installed during construction and found to have a measured sensitivity outside the listed and marked sensitivity range are to be cleaned or replaced in accordance with NFPA 72-1999 Chapter 7 at completion of construction.

NFPA 72-1999 2-3.6.2.2

Holes in the back of a detector will be covered by a gasket, sealant, or equivalent means, and the detector will be mounted so that airflow from inside or around the housing does not prevent the entry of smoke during a fire or test condition.

NFPA 72-1999 2-3.6.4.2

Air-sampling detectors will give a trouble signal where the airflow is outside the manufacturer's specified range. The sampling ports and in-line filter, if used, will be kept clear in accordance with the manufacturer's documented instructions.

NFPA 72-1999 2-3.6.4.3

Air sampling network of piping and fittings will be airtight and permanently fixed. Sampling system piping will be conspicuously identified as SMOKE DETECTOR SAMPLING TUBE. DO NOT DISTURB, as follows:

(a) At changes in direction or branches of piping
(b) At each side of penetrations of walls, floors, or similar barriers
(c) At intervals on piping sufficient to provide ready visibility within the space, but no greater than 20 feet

NFPA 72-1999 2-3.6.5

NFPA 72-1999 A-2.3.6.5

High Rack Storage. Where smoke detectors are installed to actuate a suppression system, NFPA 13, Standard for the Installation of Sprinkler Systems, will apply.

For the most effective detection of fire in high rack storage areas, detectors should be located on the ceiling above each aisle and at intermediate levels in the racks. This is necessary to detect smoke that is trapped in the racks at an early stage of fire development, when insufficient thermal energy is released to carry the smoke to the ceiling. Earliest detection of smoke is achieved by locating the intermediate level detectors adjacent to alternate pallet sections as shown in Figures A-2-3.6.5(a) and (b). The detector manufacturer's recommendations and engineering judgement should be followed for specific installations.

A projected beam-type detector is permitted to be used in lieu of a single row of individual spot-type smoke detectors.

Sampling ports of an air sampling-type detector may be permitted to be located above each aisle to provide coverage equivalent to the location of spot-type detectors. The manufacturer's recommendations and engineering judgement should be followed for the specific installation.

NFPA 72-1999 2-3.6.6.2

Smoke detectors will not be located directly in the airstream of supply registers.

NFPA 72-1999 2-3.6.6.3

NFPA 72-1999 A-2-3.6.6.3

Spacing in High Air Movement Areas. Smoke detector spacing will be in accordance with NFPA 72-1999 Table 2-3.6.6.3 and Figure 2-3.6.6.3, except air-sampling or projected beam smoke detectors installed in accordance with the manufacturer's documented instructions.

Smoke detectors spacing depends on the movement of air within the room.

33024 RADIANT ENERGY SENSING FIRE DETECTORS

REFERENCES:

NFPA 72

DESCRIPTION:

Understand the basic principles of operation of flame flicker, infrared, photoelectric and ultraviolet flame sensing fire detectors and the particular location and spacing requirements for flame detectors when used on indoor and outdoor applications.

NFPA 72-1999 1-4

Flame. A body or stream of gaseous material involved in the combustion process and emitting radiant energy at specific wavelength bands determined by the combustion chemistry of the fuel. In most cases, some portion of the emitted radiant energy is visible to the human eye.

NFPA 72-1999 1-4

NFPA 72-1999 A-1-4

Flame Detector. A radiant energy-sensing fire detector that detects the radiant energy emitted by a flame. (See NFPA 72-1999 A-2-4.2.)

Flame detectors are categorized as ultraviolet, single wavelength infrared, ultraviolet infrared, or multiple wavelength infrared.

NFPA 72-1999 1-4

NFPA 72-1999 A-1-4

Wavelength. The distance between the peaks of a sinusoidal wave. All radiant energy can be described as a wave having a wavelength. Wavelength serves as the unit of measure for distinguishing between different parts of the spectrum. Wavelengths are measured in microns (:M), or Angstroms (Å).

The concept of wavelength is extremely important in selecting the proper detector for a particular application. There is a precise interrelation between the wavelength of light being emitted from a flame and the combustion chemistry producing the flame. Specific subatomic, atomic, and molecular events yield radiant energy of specific wavelengths. For example, ultraviolet photons are emitted as the result of the complete loss of electrons or very large changes in electron energy levels. During combustion, molecules are violently torn apart by the chemical reactivity of oxygen, and electrons are released in the process, recombining at drastically lower energy levels, thus giving rise to ultraviolet rations. Visible radiation is generally the result of smaller changes in electron energy levels within the molecules of fuel, flame intermediates, and products of combustion. Infrared radiation comes from the vibration of molecules or parts of molecules when they are in the superheated state associated with combustion. Each chemical compound exhibits a group of wavelengths at which it is resonant. These wavelengths constitute the chemical's infrared spectrum, which is usually unique to that chemical.

This interrelationship between wavelength and combustion chemistry affects the relative performance of various types of detectors with respect to various fires.

NFPA 72-1999 A-2-4.1

Radiant Energy. For the purpose of NFPA 72-1999, radiant energy includes the electromagnetic radiation emitted as a by-product of the combustion reaction, which obeys the laws of optics. This includes radiation in the ultraviolet (0.1 to 0.35 μm), visible (0.36 to 0.75 μm), and infrared (0.76 to 220 μm) portions of the spectrum emitted by flames or glowing embers.

Conversion Factors: 1.0 μm = 1000 nM = 10,000 Å

NFPA 72-1999 2-4.2

NFPA 72-1999 A-2-4.2

Fire Characteristics and Detector Selection.

Following are operating principles for two types of detectors.

(a) Flame Detectors. Ultraviolet flame detectors typically use a vacuum photodiode Geiger-Muller tube to detect the ultra violet radiation that is produced by a flame. The photodiode allows a burst of current to flow for each ultraviolet photon that hits the active area of the tube. When the number of current bursts per unit time reaches a predetermined level, the detector initiates an alarm.

A single wavelength infrared flame detector uses one of several different photocell types to detect the infrared emissions in a single wavelength band that are produced by a flame. These detectors generally include provisions to minimize alarms from commonly occurring infrared sources such as incandescent lighting or sunlight.

An ultraviolet/infrared (UV/IR) flame detector senses ultraviolet radiation with a vacuum photodiode tube and a selected wavelength of infrared radiation with a photocell and uses the combined signal to indicate a fire. These detectors need exposure to both types of radiation before an alarm signal can be initiated.

A multiple wavelength infrared (IR/IR) flame detector senses radiation at two or more narrow bands of wavelengths in the infrared spectrum. These detectors electronically compare the emissions between the band and initiate a signal where the relationship between the two bands indicates a fire.

(b) Spark/Ember Detectors. A spark/ember-sensing detector usually uses a solid state photodiode or phototransistor to sense the radiant energy emitted by embers, typically between 0.5 microns and 2.0 microns in normally dark environments. These detectors can be made extremely sensitive (microwatts), and their response times can be made very short (microseconds).

NFPA 72-1999 2-4.2.1

NFPA 72-1999 A-2-4.2.1

The type and quantity of radiant energy-sensing fire detectors will be determined based upon the performance characteristics of the detector and an analysis of the hazard, including the burning characteristics of the fuel, the fire growth rate, the environment, the ambient conditions, and the capabilities of the extinguishing media and equipment.

The radiant energy from a flame or spark/ember is comprised of emissions in various bands of the ultraviolet, visible, and infrared portions of the spectrum. The relative quantities of radiation emitted in each part of the spectrum are determined by the fuel chemistry, the temperature, and the rate of combustion. The detector should be matched to the characteristics of the fire.

Almost all materials that participate in flaming combustion emit ultraviolet radiation to some degree during flaming combustion, whereas only carbon-containing fuels emit significant radiation at the 4.35 micron (carbon dioxide) band used by many detector types to detect a flame. (*See Figure A-2-4.2.1*)

The radiant energy emitted from an ember is determined primarily by the fuel temperature (Planck's Law Emissions) and the emissivity of the fuel. Radiant energy from an ember is primarily infrared and, to a lesser degree, visible in wavelength. In general, embers do not emit ultraviolet energy in significant quantities (0.1 percent of total emissions) until the ember achieves temperatures of 2000°K (1727°C or 3240°F). In most cases, the emissions are included in the band of 0.8 microns to 2.0 microns, corresponding to temperatures of approximately 750°F to 1830°F (398°C to 1000°C).

Most radiant energy detectors have some form of qualification circuitry within them that uses time to help distinguish between spurious, transient signals and legitimate fire alarms. These circuits become very important where the anticipated fire scenario and the ability of the detector to respond to that anticipated fire are considered. For example, a detector that utilizes an integration circuit or a timing circuit to respond to the flickering light from a fire might not respond well to a deflagration resulting from the ignition of accumulated combustible vapors and gases, or where the fire is a spark that's traveling up to 328 ft/sec (100 m/sec) past the detector. Under these circumstances, a detector that has a high speed response capability is most appropriate. On the other hand, in applications where the development of the fire is slower, a detector that utilizes time for the confirmation of repetitive signals is appropriate. Consequently, the fire growth rate should be considered in selecting the detector. The detector performance should be selected to respond to the anticipated fire.

The radiant emissions are not the only criteria to be considered. The medium between the anticipated fire and the detector is also very important. Different wavelengths of radiant energy are absorbed with varying degrees of efficiency by materials that are suspended in the air or that accumulate on the optical surfaces of the detector. Generally, aerosols and surface deposits reduce the sensitivity of the detector. The detection technology utilized should take into account those normally occurring aerosols and surface deposits to minimize the reduction of system response between maintenance intervals. It should be noted that the smoke evolved from the combustion of middle and heavy fraction petroleum distillates is highly absorptive in the ultraviolet end of the spectrum. Where using this type of detection, the system should be designed to minimize the effect of smoke interference on the response of the detection system.

The environment and ambient conditions anticipated in the area to be protected impact the choice of detector. All detectors have limitations on the range of ambient temperatures over which they will respond, consistent with their tested or approved sensitivities. The designer should make certain that the detector is compatible with the range of ambient temperatures anticipated in the area in which it is installed. In addition, rain, snow, and ice attenuate both ultraviolet and infrared radiation to varying degrees. Where anticipated, provisions should be made to protect the detector from accumulations of these materials on its optical surfaces.

NFPA 72-1999 2-4.2.2 The selection of the radiant energy-sensing detectors will be based on the following:

(a) Matching of the spectral response of the detector to the spectral emissions of the fire or fires to be detected,

(b) Minimizing the possibility of spurious nuisance alarms from non-fire sources inherent to the hazard area.

NFPA 72-1999 2-4.3.1.1

Radiant energy-sensing fire detectors will be employed consistent with the listing or approval and the inverse square law, which defines the fire size versus distance curve for the detector.

NFPA 72-1999 2-4.3.2

Spacing Considerations for Flame Detectors.

NFPA 72-1999 2-4.3.2.1

The location and spacing of detectors will be the result of an engineering evaluation that takes into consideration:

(a) Size of the fire that is to be detected

(b) Fuel involved

(c) Sensitivity of the detector

(d) Field of view of the detector

(e) Distance between the fire and the detector

(f) Radiant energy absorption of the atmosphere

(g) Presence of extraneous sources of radiant emissions

(h) Purpose of the detection system

(i) Response time required

NFPA 72-1999 A-2-4.3.2.1

The following are types of application for which flame detectors are suitable.

(a) High-ceiling, open-spaced buildings such as warehouses and aircraft hangers

(b) Outdoor or semi-outdoor areas where winds or draughts can prevent smoke from reaching a heat or smoke detector.

(c) Areas where rapidly developing flaming fires can occur, such as aircraft hangers, petrochemical production, storage, and transfer areas, natural gas installations, paint shops, or solvent areas

(d) Areas needing high fire risk machinery or installations, often coupled with an automatic gas extinguishing system

(e) Environments that are unsuitable for other types of detectors

Some extraneous sources of radiant emissions that have been identified as interfering with the stability of flame detectors include the following:

(a) Sunlight

(b) Lightning

(c) X-rays

(d) Gamma rays

(e) Cosmic rays

(f) Ultraviolet radiation from arc welding

(g) Electromagnetic interference (EMI, RFI)

(h) Hot objects

(i) Artificial lighting.

NFPA 72-1999 2-4.3.2.5 Because flame detectors are line-of-sight devices, special care must be taken to ensure that their ability to respond to the required area of fire in the zone that is to be protected is not compromised by the presence of intervening structural members or other opaque objects or materials.

33025 SPRINKLER WATERFLOW AND SUPERVISORY DEVICES

REFERENCES:

NFPA 13

NFPA 72

Fire Protection Handbook

*Fire Alarm
Signaling Systems*

DESCRIPTION:

Understand the operation, location, use and interconnection to a fire alarm system of various sprinkler wet and dry pipe alarm devices such as water flow, pressure, temperature and level, retarding devices, gate valve supervisory switches and fire pumps.

NFPA 72-1999 3-8.3.3.1.2

NFPA 72-1999 A-3-6.3.3.1.2

Provision will be made for supervising the conditions that are essential for the operation of sprinkler and other fire suppression systems, except those conditions related to water mains, tanks, cisterns, reservoirs, and other water supplies controlled by a municipality or a public utility.

Supervisory systems are not intended to provide indication of design, installation, or functional defects in the supervised systems or system components and are not a substitute for regular testing of those systems in accordance with the applicable standard.

Supervised conditions should include, but should not be limited to the following:

(1) Size of control valves (1 1/2 inch or larger)

(2) Pressure, including dry-pipe system air, pressure tank air, pre-action system supervisory air, steam for flooding systems, and public water

(3) Water tanks, including water level and temperature

(4) Building temperature, including areas such as valve closet and fire pump house

(5) Electric fire pumps, including running (alarm or supervisory), power failure, and phase reversal

(6) Engine-driven fire pumps, including running (alarm or supervisory), failure to start, controller of "automatic," and trouble (for example, low oil, high temperature, overspeed)

(7) Steam turbine fire pumps, including running (alarm or supervisory), steam pressure, and steam control valves

(8) Fire suppression systems appropriate to the system employed

NFPA 72-1999 2-6.2

Initiation of the alarm signal must occur within 90 seconds of waterflow at the alarm-initiating device when flow occurs that is equal to or greater than that from a single sprinkler of the smallest orifice size installed in the system. Movement of water due to waste, surges, or variable pressure must not be indicated.

NFPA 72-1999 2-6.3

Piping between the sprinkler system and a pressure actuated alarm-initiating device must be galvanized or of non-ferrous metal or other approved corrosion-resistant material of not less than 3/8-inch nominal pipe size.

NFPA 72-1999 2-9.3

Water Level Supervisory Signal-Initiating Device. Two separate and distinct signals must be initiated: one indicating that the required water level has been lowered or raised, and the other indicating restoration.

NFPA 72-1999 2-9.3.1

A pressure tank signal-initiating device must indicate both high- and low-water level conditions. A signal must be initiated when the water level falls 3 inches or rises 3 inches.

NFPA 72-1999 2-9.3.2

A supervisory signal-initiating device for other than pressure tanks must initiate a low-water level signal when the water level falls 12 inches.

NFPA 72-1999 2-9.4

Water Temperature Supervisory Signal-Initiating Device. A temperature supervisory device for a water storage container exposed to freezing conditions must initiate two separate and distinctive signals. One signal will indicate a decrease in water temperature to 40°F and the other will indicate its restoration to above 40°F.

NFPA 72-1999 2-9.5

Room Temperature Supervisory Signal-Initiating Device. A room temperature supervisory device must indicate a decrease in room temperature to 40°F, its restoration to above 40°F.

33026 ALARM NOTIFICATION APPLIANCES

REFERENCES:

NFPA 72
NFPA 101

Fire Alarm
Signaling Systems

Manufacture's Text

Americans with
Disabilities Act
Accessibility Guidelines

DESCRIPTION:

Understand the operation, selection, location, spacing, mounting and use of audible and visible alarm notification appliances.

NFPA 72-1999 1-4

Operating Mode, Private. Audible or visible signaling only to those persons directly concerned with the implementation and direction of emergency action initiation and procedure in the area protected by the fire alarm system.

NFPA 72-1999 1-4

Operating Mode, Public. Audible or visible signaling to occupants or inhabitants of the area protected by the fire alarm system.

NFPA 72-1999 1-4

Coded. An audible or visible signal that conveys several discrete bits or units of information. Notification signal examples are numbered strokes of an impact-type appliance and numbered flashes of a visible appliance.

NFPA 72-1999 1-4

Noncoded Signal. An audible or visible signal conveying one discrete bit of information.

NFPA 72-1999 1-4

Notification Appliance. A fire alarm system component such as a bell, horn, speaker, light, or text display that provides audible, tactile, or visible outputs, or any combination thereof.

> **Audible Notification Appliance.** A notification appliance that alerts by the sense of hearing.

> **Tactile Notification Appliance.** A notification appliance that alerts by the sense of touch or vibration.

> **Visible Notification Appliance.** A notification appliance that alerts by the sense of sight.

Textual Audible Notification Appliance. A notification appliance that conveys a stream of audible information. An example of a textual audible notification appliance is a speaker that reproduces a voice message.

Textual Visible Notification Appliance. A notification appliance that conveys a stream of visible information that displays an alphanumeric or pictorial message. Textual visible notification appliances provide temporary text, permanent text, or symbols. Textual visible notification appliances include, but are not limited to, annunciators, monitors, CRT's, displays, and printers.

NFPA 72-1999 4-3.1.1

An average sound level greater than 105dBA will require the use of a visible signal appliance(s) in accordance with NFPA 72-1999 Section 4-4.

NFPA 72-1999 4-3.2.1

Audible notification appliances intended for operation in the public mode must have a sound level of not less than 75 dBA at 10 ft or more than 120 dBA at the minimum hearing distance from the audible appliance.

NFPA 72-1999 4-3.2.2

To ensure that audible public mode signals are clearly heard, they must have a sound level at least 15dBA above the average ambient sound level or 5 dBA above the maximum sound level having a duration of at least 60 seconds, whichever is greater, measured 5 feet above the floor in the occupiable area.

Audible alarm notification appliances installed in elevator cars are permitted to use the audibility criteria for private mode appliances detailed in NFPA 72-1999 4-3.3.2

NFPA 72-1999 4-3.5.1

If ceiling heights allow, wall-mounted audible notification appliances must have their tops at heights above the finished floors at heights of not less than 90 inches and below the finished ceilings of not less than 6 inches. This requirement does not preclude ceiling-mounted or recessed appliances.

Different mounting heights are permitted by the authority having jurisdiction provided the sound pressure level requirements of NFPA 72-1999 4-3.2 and 4-3.3 are met.

NFPA 72-1999 4-4.3

Appliance Photometrics. The prescriptive requirements of Section 4-4 assume the use of appliances having very specific characteristics of light, color, intensity, distribution, and so on. The appliance and application requirements are based on extensive research. However, the research was limited to typical residential and commercial applications such as school classrooms, offices, hallways, and hotel rooms. While these specific appliances and applications will likely work in other spaces, their use might not be the most effective solution and might not be as reliable as other visible notification methods.

For example, in large warehouse spaces and large distribution spaces such as super stores, it is possible to provide visible signaling using the appliances and applications of this chapter. However, mounting strobe lights at a height of 80 inches to 96 inches along aisles with rack storage subjects the lights to frequent mechanical damage by forklift trucks and stock. Also, the number of appliances required would be very high. It might be possible to use other appliances and applications not specifically addressed by this chapter at this time. Alternative applications must be carefully engineered for reliability and function and would require permission of the authority having jurisdiction.

Visible notification using the methods contained in NFPA 72-1999 4-4.4.1 is achieved by indirect signaling. This means the viewer need not actually see the appliance, just the effect of the appliance. This can be achieved by producing minimum illumination on surfaces near the appliances such as the floor, walls, and desks. There must be a sufficient change in illumination to be noticeable. The tables and charts in Section 4-4 specify a certain candela effective light intensity for certain size spaces. The data were based on extensive research and testing. Appliances do not typically produce the same light intensity when measured off-axis. To ensure that the appliance produces the desired illumination (effect), it must have some distribution of light intensity to the areas surrounding the appliance. UL 1971, *Standard for Safety Signaling Devices for the Hearing Impaired,* specifies the distribution of light shown to provide effective notification by indirect visible signaling.

NFPA 72-1999 4-4.4

Appliance Location. Wall-mounted visible notification appliances must be mounted such that the entire lens is not less than 80 inches and not greater than 96 inches above the finished floor.

NFPA 72-1999 4-4.4.2.1 NFPA 72-1999 Table 4-4.4.2.1 applies to corridors not exceeding 20 feet in width. For corridors greater than 20 feet wide, Figure 4-4.4.1.1 and Tables 4-4.4.1.1(a) and (b) apply. In a corridor application, visible appliances must be rated not less than 15 cd.

NFPA 72-1999 4-4.4.2.2 Visible appliances in corridors must be located no more than 15 feet from the end of the corridor with a separation not greater than 100 feet between appliances. If there is an interruption of the concentrated viewing path, such as a fire door, an elevation change, or any other obstruction, the area will be treated as a separate corridor.

NFPA 72-1999 4-4.4.2.1

33027 BASICS OF SIGNAL TRANSMISSION

<table>
<tr><td>REFERENCES:</td><td>DESCRIPTION:</td></tr>
<tr><td>NFPA 72

Related Appendix</td><td>Know the types of signaling circuits on a fire alarm system, the types of signals transmitted over these circuits, and the limitations on the capacity of the signaling circuits.</td></tr>
</table>

NFPA 72-1999 1-4

Active Multiplex System. A multiplexing system in which signaling devices such as transponders are employed to transmit status signals of each initiating device or initiating device circuit within a prescribed time interval so that lack of receipt of such signal may be interpreted as a trouble signal.

NFPA 72-1999 1-4

Multiplexing. A signaling method characterized by simultaneous or sequential transmission, or both, and reception of multiple signals on a signaling line circuit, a transmission channel, or a communications channel, including means for positively identifying each signal.

NFPA 72-1999 5-3

Proprietary Central Station Systems.

NFPA 72-1999 5-3.4.7

The maximum elapsed time from sensing a fire alarm at an initiating device or initiating device circuit until it is recorded or displayed at the proprietary supervising station must not exceed 90 seconds.

NFPA 72-1999 5-5

Communications Methods for Supervising Station Fire Alarm Systems.

NFPA 72-1999 5-5.2.2.1

Equipment. Fire alarm system equipment and installations must comply with Federal Communication Commission (FCC) rules and regulations, as applicable, concerning the following:

(1) Electromagnetic radiation

(2) Use of radio frequencies

(3) Connection to the public switched telephone network of telephone equipment systems, and protection apparatus.

NFPA 72-1999 5-5.3.1.2.3

NFPA 72-1999 A-5-5.3.1.2.3(b)

The maximum end-to-end operating time parameters allowed for an active multiplex system are as follows:

(a) The maximum allowable time lapse from the initiation of a single fire alarm signal until it is recorded at the supervising station must not exceed 90 seconds. When any number of subsequent fire alarm signals occur at any rate, they will be recorded at a rate no slower than one every 10 additional seconds.

(b) The maximum allowable time lapse from the occurrence of an adverse condition in any transmission channel until recording of the adverse condition is started must not exceed 90 seconds for Type 1 and Type 2 systems, and 200 seconds for Type 3 systems. The requirements of NFPA 72-1999 5-5.3.1.3 apply.

Derived channel systems comprise Type 1 and Type 2 systems only.

(c) In addition to the maximum operating time allowed for fire alarm signals, the requirements of one of the following must be met:

1. A system unit that has more than 500 initiating device circuits must be able to record not less than 50 simultaneous status changes in 90 seconds, or

2. A system unit having fewer than 500 initiating device circuits must be able to record not less than 10 percent of the total number of simultaneous status changes within 90 seconds.

Except proprietary supervising station systems that have opening time requirements specified in NFPA 72-1999 5-3.4.7 through 5-3.4.9.

NFPA 72-1999 5-5.3.1.3

System Classification. Active multiplex systems are divided into three categories based on their ability to perform under adverse conditions of their transmission channels. The system classifications are as follows:

(a) A Type 1 system will have dual control as described in NFPA 72-1999 5-5.2.4. An adverse condition on a trunk or leg facility must not prevent the transmission of signals from any other trunk or leg facility, except those signals normally dependent on the portion of the transmission channel in which the adverse condition has occurred. An adverse condition limited to a leg facility will not interrupt normal A service on any trunk or other leg facility. The requirements of NFPA 72-19999 5-5.2.1, 5-5.2.2, and 5-5.2.3 must be met by Type 1 systems.

(b) A Type 2 system will have the same requirements as a Type 1 system except that dual control of the primary trunk facility is not required.

(c) A Type 3 system will automatically indicate and record at the supervising station the occurrence of an adverse condition on the transmission channel between a protected premises and the supervising station. The requirements of NFPA 72-1999 5-5.2 will be met.

The requirements of NFPA 72-1999 5-5.2.4 do not apply.

NFPA 72-1999 5-5.3.1.4 **System Loading Capacities.** The capacities of active multiplex systems are based on the overall reliability of the signal receiving, processing, display, and recording equipment at the supervising and subsidiary stations, and have the capability to transmit signals during adverse conditions of the signal transmission facilities. Allowable capacities of active multiplex systems must be in accordance with NFPA 72-1999 Table 5-5.3.1.4.

NFPA 72-1999 5-5.3.2.1.1 **Digital Alarm Communicator Transmitter (DACT).** A DACT must be connected to the public switched telephone network upstream of any private telephone system at the protected premises. In addition, the connections to the public switched telephone network will be under the control of the subscriber for whom service is being provided by the supervising station fire alarm system. Special attention is required to ensure that this connection is made only to a loop start telephone circuit and not to a ground start telephone circuit.

If public cellular telephone service is utilized as a secondary means of transmission, the requirements of NFPA 72-1999 5-5.3.2.1.1 does not apply to the cellular telephone service.

NFPA 72-1999 5-5.3.2.1.2 All information exchanged between the DACT at the protected premises and the digital alarm communicator receiver (DACR) at the supervising or subsidiary station will be by digital code or some other approved means. Signal repetition, digital parity check, or some other approved means of signal verification will be used.

NFPA 72-1999 5-5.3.2.1.3 A DACT will be configured so that when it is required to transmit a signal to the supervising station, it will seize the telephone line (going of-hook) at the protected premises, disconnect an outgoing or incoming telephone call, and prevent use of the telephone line for outgoing telephone calls until signal transmission has been completed. A DACT must not be connected to a party line telephone facility.

NFPA 72-1999 5-5.3.2.1.4 A DACT will have the means to satisfactorily obtain a dial tone, dial the number(s) of the DACR, obtain verification that the DACR is ready to receive signals, transmit the signal, and receive acknowledgment that the DACR has accepted that signal. In no event will the time from going off-hook exceed 90 seconds per attempt.

NFPA 72-1999 5-5.3.2.1.5 A DACT will have suitable means to reset and retry where the first attempt to complete a signal transmission sequence is unsuccessful. A failure to complete a connection must not prevent subsequent attempts to transmit an alarm where such alarm is generated from any other initiating device circuit or signaling line circuit, or both. Additional attempts will be made, until the signal transmission sequence has been completed, up to a minimum of 5 and a maximum of 10 attempts.

If the maximum number of attempts to complete the sequence is reached, an indication of the failure will be made at the premises.

NFPA 72-1999 5-5.3.2.1.6.1

NFPA 72-1999 5-5.3.2.1.6(8)

A DACT will employ one of the following combinations of transmission channels:
    (a) Two telephone lines (numbers), or
    (b) One telephone line (number) and one cellular telephone connection, or
    (c) One telephone line (number) and a one-way radio system, or
    (d) One telephone line (number) equipped with a derived local channel, or
    (e) One telephone line (number) and a one-way private radio alarm system, or
    (f) One telephone line (number) and a private microwave radio system, or
    (g) One telephone line (number) and a two-way RF multiplex system, or

(h) A single integrated services digital network (ISDN) telephone line using a terminal adapter specifically listed for supervising station fire alarm service, where the path between the transmitter and the switched telephone network serving central office is monitored for integrity so that the occurrence of an adverse condition in the path will be annunciated at the supervising station within 200 seconds.

A two number ISDN line is not a substitute for the requirement to monitor the integrity of the path.

NFPA 72-1999 5-5.3.2.1.6.2 The following requirements apply to all combinations in NFPA 72-1999 5-5.3.2.1.6.1:

(a) Both channels will be supervised in a manner appropriate for the means of transmission employed., and
(b) Both channels will be tested at intervals not exceeding 24 hours

For public cellular telephone service, a verification (test) signal will be transmitted at least monthly.

Where two telephone lines (numbers) are used, it is permitted to test each telephone line (number) at alternating 24-hour intervals.

(c) The failure of either channel must send a trouble signal on the other channel within 4 minutes.
(d) When one transmission channel has failed, all status change signals will be sent over the other channel.

Where used in combination with a DACT, a derived local channel is not required to send status change signals other than those indicating that adverse conditions exist on the telephone line (number).

(e) The primary channel will be capable of delivering an indication to the DACT that the supervising station has received the message.
(f) The first attempt to send a status change signal must use the primary channel except where the primary channel is known to have failed.
(g) Simultaneous transmission over both channels is permitted.
(h) Failure of telephone lines (numbers) or cellular service will be annunciated locally.

NFPA 72-1999 5-5.3.2.1.7.1 — A DACT will be connected to two separate means of transmission at the protected premises. The DACT must be capable of selecting the operable means of transmission in the event of failure of the other means. The primary means of transmission will be a telephone line (number) connected to the public switched network.

NFPA 72-1999 5-5.3.2.1.7.2 — The first transmission attempt will utilize the primary means of transmission.

NFPA 72-1999 5-5.3.2.1.8 — Each DACT will be programmed to call a second DACR line (number) where the signal transmission sequence to the first called line (number) is unsuccessful.

NFPA 72-1999 5-5.3.2.1.9 — If long distance telephone service, including WATS, is used, the second telephone number must be provided by a different long distance service provider if there are multiple providers.

NFPA 72-1999 5-5.3.2.1.10 — Each DACT will automatically initiate and complete a test signal transmission sequence to its associated DACR at least once every 24 hours. A successful signal transmission sequence of any other type within the same 24-hour period is considered sufficient to fulfill the requirement to verify the integrity of the reporting system, provided signal processing is automated so that 24-hour delinquencies are individually acknowledged by supervising station personnel.

NFPA 72-1999 5-5.3.2.1.11 — If a DACT is programmed to call a telephone line (number) that is call forwarded to the line (number) of the DACR, it must will be implemented to verify the integrity of the call forwarding feature every 4 hours.

NFPA 72-1999 5-5.3.2.2.1.1 — Spare DACRs must be provided in the supervising or subsidiary station. The spare DACRs will be able to be switched into the place of a failed unit within 30 seconds after detection of failure.

One spare DACR is permitted to serve as a backup for up to five DACRs in use.

NFPA 72-1999 5-5.3.2.2.1.2 — The number of incoming telephone lines to a DACR is limited to eight lines.

If the signal receiving, processing, display, and recording equipment at the supervising or subsidiary station is duplicated and a switchover is able to be accomplished in less than 30 seconds with not loss of signal during this period, the number of incoming lines to the unit is be permitted to be unlimited.

NFPA 72-1999 5-5.3.2.2.2.1 The DACR equipment at the supervising or subsidiary station must be connected to a minimum of two separate incoming telephone lines (numbers). If the lines (numbers) are in a single hunt group, they will be individually accessible; otherwise, separate hunt groups are required. These lines (numbers) will be used for no other purpose than receiving signals from a DACT. These lines (numbers) must be unlisted.

NFPA 72-1999 5-5.3.2.2.2.2 The failure of any telephone line (number) connected to a DACR due to loss of line voltage must be annunciated visually and audibly in the supervising station.

NFPA 72-1999 5-5.3.2.2.2.3 The loading capacity for a hunt group will be in accordance with Table NFPA 72-1999 5-5.3.2.2.2.3 or be capable of demonstrating a 90 percent probability of immediately answering an incoming call.

Each supervised intrusion alarm (open/close) or each suppressed guard tour transmitter will reduce the allowable DACTs as follows:

1. Up to a four-line hunt group, by 10
2. Up to a five-line hunt group, by 7
3. Up to a six-line hunt group, by 6
4. Up to a seven-line hunt group, by 5
5. Up to a eight-line hunt group, by 4

Each guard tour transmitter will reduce the allowable DACTs as follows:

1. Up to a four-line hunt group, by 30
2. Up to a five-line hunt group, by 21
3. Up to a six-line hunt group, by 18
4. Up to a seven-line hunt group, by 15
5. Up to a eight-line hunt group, by 12

NFPA 72-1999 5-5.3.2.3 **Digital Alarm Radio System (DARS).**
If any DACT signal transmission is unsuccessful, the information must
NFPA 72-1999 5-5.3.2.3.1 be transmitted by means of the digital alarm radio transmitter (DART). The DACT will continue its normal transmission sequence as required by NFPA 72-1999 5-5.3.2.1.5.

NFPA 72-1999 5-5.3.2.3.2 — The DARS must be capable of demonstrating a minimum of 90 percent probability of successfully completing each transmission sequence.

NFPA 72-1999 5-5.3.2.3.3 — Transmission sequences must be repeated a minimum of five times. The digital alarm radio transmitter (DART) transmission is permitted to be terminated in less than five sequences if the DACT successfully communicates to the DACR.

NFPA 72-1999 5-5.3.2.3.4 — Each DART will automatically initiate and complete a test signal transmission sequence to its associated digital alarm radio receiver (DARR) at least once every 24 hours. A successful DART signal transmission sequence of any other type within the same 24-hour period is considered sufficient to fulfill the requirement to test the integrity of the reporting system, provided signal processing is automated so that 24-hour delinquencies are individually acknowledged by supervising station personnel.

NFPA 72-1999 5-5.3.2.4 — **Digital Alarm Radio Transmitter (DART).** A DART will transmit a digital code or another approved signal by use of radio transmission to its associated digital alarm radio receiver (DARR). Signal repetition, digital parity check, or an equivalent means of signal verification will be used. The DART must comply with applicable FCC rules consistent with its operating frequency.

NFPA 72-1999 5-5.3.2.5
NFPA 72-1999 5-5.3.2.5.1.1 — **Digital Alarm Radio Receiver (DARR).**
A spare DARR must be provided in the supervising station and be able to be switched into the place of a failed unit within 30 seconds after detection of failure.

NFPA 72-1999 5-5.3.2.5.2 — Facilities must be provided at the supervising station for supervisory and control functions of subsidiary and repeater station radio receiving equipment. This will be accomplished via a supervised circuit where the radio equipment is remotely located from the supervising or subsidiary station. The following conditions will be supervised at the supervising station:

    (a) Failure of ac power supplying the radio equipment
    (b) Malfunction of the receiver
    (c) Malfunction of the antenna and interconnecting cable
    (d) Indication of automatic switchover of the DARR
    (e) Malfunction of data transmission line between the DARR
        and the supervising or subsidiary station

NFPA 72-1999

McCulloh Systems.

HANDBOOK 5-5.3.3

McCulloh systems provide the oldest form of transmission between a protected premises and a supervising station. Coded transmitters at a protected premises connect in series with transmitters at other protected premises and with receiving equipment at the supervising station. The interconnected circuits must maintain continuous metallic circuit continuity. This allows the dc current to flow from the power supply at the supervising station, out over the series circuit, and through the coded contacts of the transmitters at the protected premises. Where the public telephone utility does not wish to maintain circuits that will offer continuous metallic circuit continuity, an alternative exists (see NFPA 72-1999 5-5.3.2.6). This alternative system converts the McCulloh-type system into a multiplex system between public telephone utility company wire centers. Then it reconverts the multiplex system to a McCulloh system in order to deliver the signals to the supervising station.

Initiation of an alarm, supervisory, or trouble signal at the protected premises actuates the associated transmitter. As the code wheel of the actuated transmitter turns, it alternately breaks the circuit and connects the circuit to earth ground.

Under normal circumstances, the breaking of the circuit operates receiving equipment that records the coded pulses at the supervising station. Operators, either manually or by a computer-based automation system at the supervising station, convert these pulses to information that gives the location of the protected premises.

If a single open fault or single ground fault impairs the circuit between the protected premises and the supervising station, then the signal produced by the turning of the code wheel transmits through earth ground. Operators must respond to a trouble signal generated by the fault. They then manually, or by a computer-based automation system, recondition the circuit to receive the signals through earth ground.

NFPA 72-1999 5-5.3.3.1

Transmitters.

NFPA 72-1999 5-5.3.3.3.1

A coded alarm signal from a transmitter must consist of not less than three complete rounds of the number or code transmitted.

NFPA 72-1999 5-5.3.3.1.2

A coded fire alarm box will produce not less than three signal impulses for each revolution of the coded signal wheel or other approved device.

NFPA 72-1999 5-5.3.3.3	**Loading Capacity of McCulloh Circuits.**
NFPA 72-1999 5-5.3.3.3.1	The number of transmitters connected to any transmission channel is limited to avoid interference. The total number of code wheels or other approved devices connected to a single transmission channel must not exceed 250. Alarm signal transmission channels are reserved exclusively for fire alarm signal transmitting service, except as provided in NFPA 72-1999 5-5.3.3.3.4.
NFPA 72-1999 5-5.3.3.3.2	The number of waterflow switches permitted to be connected to actuate a single transmitter must not exceed five switches.
NFPA 72-1999 5-5.3.3.3.3	The number of supervisory switches permitted to be connected to actuate a single transmitter must not exceed 20 switches.
NFPA 72-1999 5-5.3.3.3.6	One alarm transmission channel must not serve more than 25 plants. A plant is permitted to consist of one or more buildings under the same ownership, and the circuit arrangement will be such that an alarm signal cannot be received from more than one transmitter at a time within a plant. If such noninterference is not provided, each building will be considered a plant.
NFPA 72-1999 5-5.3.3.3.7	One sprinkler supervisory transmission channel circuit may not serve more than 25 plants. A plant is permitted to consist of one or more buildings under the same ownership.

Two-Way Radio Frequency (RF) Multiplex Systems.

33028 BUSINESS COMMUNICATIONS

REFERENCES:

See basic grammar and writing handbooks

DESCRIPTION:

Use the rules of syntax and style to write clear sentences and paragraphs in preparing routine correspondence and reports. Follow standard business communications procedures.

For productive letter writing:

- Plan your communication as a whole before you start to write.

- Do not use stilted or worn-out expressions. Use clear and straightforward language.

- Do not use unnecessary word or phrases.

- Avoid using favorite words and expressions.

- Breakup overlong, stuffy sentences by making short sentences of dependent clauses.

- Use different words to express various shades of meaning. (Example: Use aroma, fragrance, scent, or odor, instead of smell.)

- Never use a word that might humiliate or belittle the reader.

- Avoid all forms of sexist language.

- Eliminate language that reveals or focuses on racial or ethnic characteristics.

- Do not use words and phrases that are demeaning.

The opening of any letter must get the reader's attention immediately. Seven suggestions that will help you develop a technique of starting a letter are:

1) Make the opening short

2) Go straight to the point

3) Avoid stilted openings

4) Include the reader's name

5) Refer to the previous contact

6) Use a pleasant phrase

7) Use the "Who, What, When, Where, Why, and How" technique.

The closing of a letter should add something definite to the letter or it should not be there. Four techniques to help you write a closing that will add something to your letters are:

1) Avoid stilted or formal endings

2) Suggest only one action

3) Use dated action

4) Use positive words

33029 INTERMEDIATE MATHEMATICS

REFERENCES:

Seebasic textbooks on algebra and trigonometry

DESCRIPTION:

Perform mathematical calculations utilizing basic algebra (fundamental laws, algebraic expressions), geometry, and the trigonometric functions of triangles.

Know how to solve the following problems:

Decimals and Percentages

What is the percentage of 33 to 132?   Divide 33 by 132 = .25 or 25%

What percentage is 42 of 16?   Divide 42 by 16 = 2.62 multiply by 100 = 262%

Gas prices decreased from $1.30 to $1.15 per gallon.
What was the percent decrease in price?

$$\frac{1.30 - 1.15}{1.30} \times 100 = 11.54\% \text{ decrease}$$

Linear Equations

Linear equations involving one variable are solved by calculating the value of the variable in one equation and then substituting this value in the other equation.

Example $2x + 4 = 6x - 12$ find x
$$6x - 2x = 4 + 12$$
$$4x = 16$$
$$x = 4$$

Now substitute this value into the 2nd equation of: $1 - x = ?$
$$1 - 4 = -3$$

Simultaneous Equations

To solve simultaneous equations with 2 unknowns, solve one equation in terms of one unknown and substitute this value into the other equation.

Solve the two following equations: $2x + y = 25$
$$x - 2y = -5$$

First: $x - 2y = -5$ Now substitute: $2x + y = 25$
$$x = -5 + 2y$$
$$2(-5 + 2y) + y = 25$$
$$-10 + 4y + y = 25$$
$$5y - 10 = 25$$
$$5y = 35$$
$$y = 7$$

Now substitute 7 for y to get the answer for x:

$$x - 2y = -5$$
$$x - 2(7) = -5$$
$$x - 14 = -5$$
$$x = 9$$

Areas and Volumes

ITEM	AREA (PLAIN)	VOLUME	AREA (3D)
Triangle	$A = bh/2$		
Square	$A = b^2$	$V = b^3$	$A = b^2 \times 6$
Recatangle	$A = ab$	$V = abc$	$A = 2(ab + bc + ac)$
Circle	$A = \pi r^2$	$V = 4/3\ \pi r^3$	$A = 4\pi r^2$

Trigonometry

Pythagorean theorem is a basic way of finding what an unknown side of a right triangle is by using the other two sides. The angle opposite the right angle is always known a c.

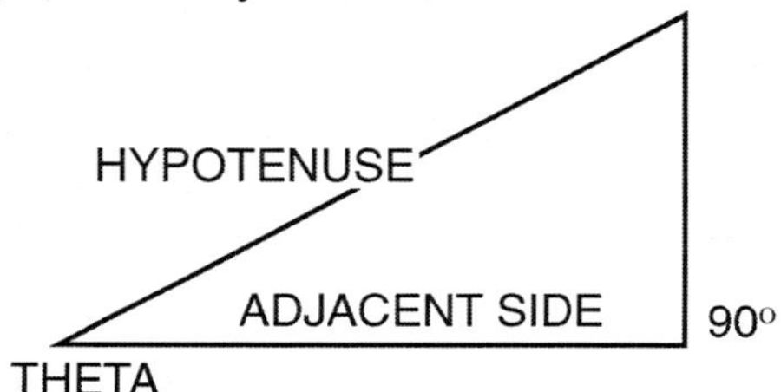

Pythagoreans theory is: $a^2 + b^2 = c^2$

Now lets take a triangle with the two adjoining sides to the right angle being 3 and 4. We need to know what the other side is.

$$3^2 + 4^2 = c^2$$
$$9 + 16 = c^2$$
$$25 = c^2$$
$$5 = c$$

Trigonometric functions are calculated using a simple set of circle formulas similar to the Ohm's Law circle. Using these formulas to get a base answer you plug the base answer into the trigonometric ratios table and you will get the answer to your problem. These three trig circles are:

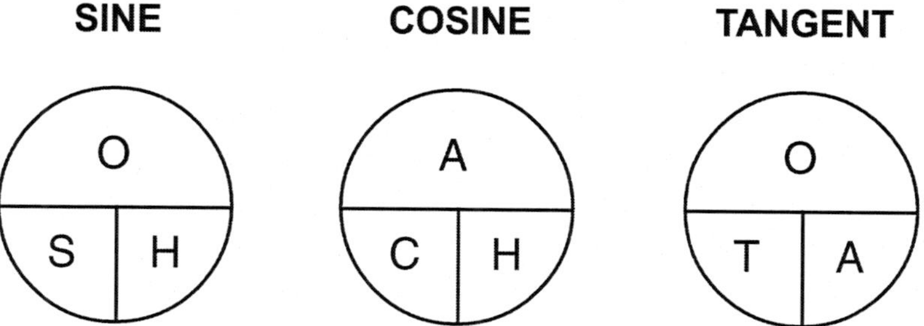

The Trigonometric Functions Chart is found on Level II Tab.

Lets take a simple problem of a triangle with the following properties. Side a is 20 feet, the angle opposite side b is 40°. What is the hypotenuse?

Because we have the length of side a and the angle opposite of side b we can use the cosine circle to find the answer.

$$cos\ 40 = 20/hyp$$

$$.766\ =\ 20/hyp$$

$$hyp = 20/.766$$

$$hyp\ =\ 26.1\ feet$$

PRACTICAL PROBLEMS

Decimals: 1. $6.774 \times 0.001\ =$

        2. $4.986 \times 4.62\ =$

        3. $376.2 \times .0022\ =$

Linear Equations: $6x + 3 - 4x\ =\ 13 - 3x$

        $4x - 1\ =\ ?$

Simultaneous Equations: $3x + 2y\ =\ 16$

        $x + 4y\ =\ 12$

Percentages: 1.   14 is what % of 50

        2.   17 is what % of 29

        3.   26 is what % of 16

Percent Increase or Decrease:

1. Halon 1301 increases in price from $6.00/lb to $9.00lb. What is the percent increase in price?

2. Halon 1301 decreases in price from $9.00/lb to $6.00lb. What is the percent decrease in price?

3. Pipe increases in price from $0.84/ft to $0.96/ft. What is the percent increase?

4. Pipe decreases in price from $11.67/10 feet to $0.92/ft. What is the percent decrease in price?

Areas & Volumes:

1. All the questions (a through d) based on a rectangle with sides of 10 and 200 feet.

   a. What is the area of the rectangle?

   b. The area of the bottm of a tank is 2,000 feet

   What is the volume of the tank if it is 12 feet deep?

   c. How much air will be left in the tank if we add 32" of water to the tank?

   d. How much air is left in the tank if we add another 18" of water?

2. What is the area of a triangle with a base of 16 feet and a height of 9 feet?

3. What is the area of a triangle with a base of 5 feet and height of 12 feet?

4. What is the volume of a room that measures 12 feet by 30 feet and has walls that are 15 feet high, the ceiling is open exposed beam that measures 12 feet in the center above the side walls.

5. What is the volume of a shed type building that measures 18 feet long by 6 feet wide, one side wall is 12 feet high the other is 22 feet high.

Area & Volume (circles and cylinders)

1.   What is the area of a circle with a radius of 9 feet?

2.   What is the area of a cross section of pipe that has a outside radius of 8 inches and an inside radius of 6 inches?

3.   What is the weight of a pipe which is 32 feet long and has the same cross sectional area as problem 2 above, if steel weighs 480 pounds per cubic foot?

4.   Determine the weight of 50 feet of pipe with an outside radius of 8 inches and a wall thickness of 1/2 inch.? (steel weighs 480 pounds per cubic foot)

Trigonometry

1.   You have a right triangle with two sides of 16 and 26 feet. What is the length of the third side. The third side is not opposite the right angle.

2.   A right triangle with a θ angle of 40° and an adjacent side of 20 feet will have a hypotenuse of ______ feet.

3.   A right triangle with a θ angle of 40° and a opposite side of 18 feet will have an adjacent side length of ______ feet.

4.   A right triangle with sides a and b being 16 and 22 feet respec tively will have a θ angle of ______ .

5.   A right triangle with a adjacent side of 29 feet and a hypontnuse of 38 feet would have a θ angle of ______ .

ANSWERS

Decimals: 1.   0.006774

2.   230.4

3.   0.8276

Linear Equation: 1.   7

Simultaneous Equations: 1. $x = 4$

2. $y = 2$

Percentages: 1.   28%

2.   59%

3.   162.5%

Percent Increase or Decrease: 1.   50%

2.   33.3%

3.   14.3%

4.   21.2%

Areas & Volumes: 1.a.   2,000 ft^2

b.   24,000 ft^3

c.   18,660 ft^3

d.   15,660 ft^3

Areas: 2.   72 ft^2

3.   30 ft^2

Volumes: 4.   7560 ft^3

5.   1836 ft^3

Area & Volume (circles & cylinders):   1.   254.5 ft^2

2.   88 in^2

3.   9387 lb

4.   4056 lb

Trigonometry: 1. $b = 20.5$ ft

2.   Hyp $= 26.2$ ft

3. $a = 21.45$ ft

4. $\theta = 54°$

5. $\theta = 40.3°$

34001 EMERGENCY VOICE/ALARM COMMUNICATION SYSTEMS

REFERENCES:

NFPA 70

NFPA 101

DESCRIPTION:

Understand an emergency voice/alarm communication system, including its purpose, operation, basic requirements, power supply sources, survivability and types of voice communication services available.

NFPA 72-1999 3-8.4.1.3

Emergency Voice/Alarm Communications. Emergency voice/alarm communications service will be provided by a system with automatic or manual voice capability that is installed to provide voice instructions to the building occupants where it is intended that there be only partial or selective evacuation or directed relocation of building occupants in the event of a fire. If emergency voice/alarm communications are used to automatically and simultaneously notify all occupants to evacuate the protected premises during a fire emergency, manual or selective paging will not be required, but, if provided, is to meet the requirements of NFPA 72-1999 Section 3-8.4.1.3

NFPA 72-1999 3-8.4.1.3.1

Application. NFPA 72-1999 3-8.4.1.3 describes the requirements for emergency voice/alarm communications. The primary purpose is to provide dedicated manual and automatic facilities for the origination, control, and transmission of information and instructions pertaining to a fire alarm emergency to the occupants (including fire department personnel) of the building. It is the intent of this section to establish the minimum requirements for emergency voice/alarm communications.

NFPA 72-1999 3-8.4.1.3.3.2

The fire command center and the central control unit is to be located within a minimum 1-hour rated fire resistive area and will have a minimum 3-foot from the front of the fire command center control equipment except, if approved by the authority having jurisdiction, the fire command center control equipment is permitted to be located in a lobby or other approved space.

NFPA 72-1999 3-8.4.1.3.3.3   If the fire command center control equipment is remote from the central control equipment, the following requirements apply:

(a) The interconnecting wiring is to be provided with mechanical protection by installing the wiring in metal conduit or metal raceway;

(b) The interconnecting wiring will be provided with resistance to attack from a fire by routing the wiring through areas whose characteristics are at least equal to the limited-combustible characteristics defined in NFPA 90A, *Standard for the Installation of Air Conditioning and Ventilating Systems*;

(c) If the interconnecting wiring exceeds 100 feet, additional resistance to attack from a fire will be provided by either of the following:

1. Installing the wiring in metal conduit or metal raceway in a 2-hour fire rated enclosure, or
2. Enclosing the wiring in a 2-hour fire rated cable assembly and installing the cable in metal conduit or metal raceway.

NFPA 72-1999 3-8.4.1.3.4.2   The secondary (standby) power supply is to be provided in accordance with NFPA 72-1999 1-5.2.5 (**Light and Power Service**) where emergency voice/alarm communications are used to notify all occupants automatically and simultaneously to evacuate the protected premises during a fire emergency. In meeting the requirements of NFPA 72-1999 1-5.2.5 (**Light and Power Service**), the secondary supply will be required to be capable of operating the system during a fire or other emergency condition for a period of 5 minutes rather than 2 hours.

NFPA 72-1999 1-5.2.6   **Secondary Supply Capacity and Sources.** The secondary supply will automatically supply the energy to the system within 30 seconds, and without loss of signals, wherever the primary supply is incapable of providing the minimum voltage required for proper operation. The secondary (standby) power supply will supply energy to the system in the event of total failure of the primary (main) power supply or when the primary voltage drops to a level insufficient to maintain functionality of the control equipment and system components. Under maximum normal load, the secondary supply will have sufficient capacity to operate a protected premises, central station, or proprietary system for 24 hours, or an auxiliary or remote station system for 60 hours; and, at the end of that period, will be capable of operating all alarm notification appliances used for evacuation or to direct aid to the location of an emergency for 5 minutes. **The secondary power supply for emergency voice/alarm communications service will be capable of operating the system under maximum normal load for 24 hours and then be capable of operating the system during a fire, or other emergency condition for a period of 2 hours. Fifteen minutes of evacuation alarm operation at maximum connected load is considered the equivalent of 2 hours of emergency operation.**

NFPA 72-1999 3-8.4.1.3.5

Voice/Alarm Signaling Service.

NFPA 72-1999 3-8.4.1.3.5.1

General. The purpose of the voice/alarm signaling service is to provide an automatic response to the receipt of a signal indicative of a fire emergency. Subsequent manual control capability of the transmission and audible reproduction of evacuation tone signals, alert tone signals, and voice directions on a selective and all-call basis, as determined by the authority having jurisdiction, is also required from the fire command center.

Where the fire command center or remote monitoring location is constantly attended by trained operators, and operator acknowledgement of receipt of a fire alarm signal is received within 30 seconds, automatic response is not required.

Where emergency voice/alarm communications are used to notify all occupants automatically and simultaneously to evacuate the protected premises during a fire emergency, the ability to give voice directions on a selective basis will not be required but, where provided, will meet the requirements of NFPA 72-1999 3-8.4.1.3.

NFPA 72-1999 3-8.4.1.3.5.2

Multichannel Capability. If required by the authority having jurisdiction, the system will allow the application of an evacuation signal to one or more zones and, at the same time, allow voice paging to the other zones selectively or in any combination.

NFPA 72-1999 3-8.4.1.3.5.3

NFPA 72-1999 3-8.4.1.3.5.3.1

Functional Sequence. In response to an initiating signal indicative of a fire emergency, the system will automatically transmit the following either immediately or after a delay acceptable to the authority having jurisdiction:

(a) If the emergency voice/alarm communications service is used to transmit a voice evacuation message, the voice message will be preceded and followed by a minimum of two cycles of the audible emergency evacuation signal specifiec in NFPA 72-1999 3-8.4.1.2 (**Distinctive Evacuation Signal).**

> (b) If the emergency voice/alarm communications service is used to transmit relocation instructions or other non-evacuation messages, a continuous alert tone of 3-second to 10-second duration followed by a message (or messages where multichannel capability is provided) will be repeated at least three times to direct the occupants of the alarm signal initiation zone and other zones in accordance with the building's fire evacuation plan.
>
> (c) An evacuation signal is to be transmitted to the alarm signal initiation zone and other zones in accordance with the building's fire evacuation plan.

*If emergency voice/alarm communications are used to notify all occupants automatically and simultaneously to evacuate the protected premises during a fire emergency, and the functional sequence described in NFPA 72-1999 3-8.4.1.3.5.3.1(a) is provided, the capability to notify portions of the protected premises selectively, is not required but, if provided, will meet the requirements of NFPA 72-1999 3-8.4.1.3 (**Emergency Voice/Alarm Communications**).*

NFPA 72-1999 3-8.4.1.3.5.3.2

Failure of the message described by NFPA 72-1999 3-8.4.1.3.5.3.1(a), if used, will sound the evacuation signal automatically. Provisions for manual initiation of voice instructions or evacuation signal generation is to be provided.

Other functional sequences are permitted where approved by the authority having jurisdiction.

If emergency voice/alarm communications are used to notify all occupants automatically and simultaneously to evacuate the protected premises during a fire emergency, provision for manual initiation of voice instructions is not required but, if provided, will meet the requirements of NFPA 72-1999 3-8.4.1.3 (**Emergency Voice/Alarm Communications**).

NFPA 72-1999 3-8.4.1.3.5.5.1

A fire command center will be provided at a building entrance or other location approved by the authority having jurisdiction. The fire command center will provide a communications center for the arriving fire department and will provide for control and display of the status of detection, alarm, and communications systems. The fire command center is permitted to be physically combined with other building operations and security centers as permitted by the authority having jurisdiction. Operating controls for use by the fire department will be clearly marked.

NFPA 72-1999 3-8.4.1.3.5.6.2 There are to be at least two loudspeakers in each paging zone of the building. Each of the loudspeakers will meet the requirements of NFPA 72-1999 Chapter 4 (**Notification Appliances for Fire Alarm Systems**).

NFPA 72-1999 3-8.4.1.3.5.6.3 Each elevator car is to be equipped with a single loudspeaker connected to the paging zone that serves the elevator group in which the elevator car is located.

NFPA 72-1999 3-8.4.1.3.6.2 If multiple notification appliance circuits are provided within an single evacuation signaling zone, all of the notification appliances within the zone are to be arranged to activate simultaneously, either automatically or by actuation of a common, manual control except where the different notification appliance circuits within an evacuation signaling zone perform separate functions, for example, presignal and general alarm signals, predischarge and discharge signals.

NFPA 72-1999 3-8.4.1.3.7.3 Two-way telephone communications service will be capable of permitting the simultaneous operation of any five telephone stations in a common talk mode.

NFPA 72-1999 3-8.4.1.3.7.5 A means for silencing the audible call-in signal-sounding appliance is permitted, provided it is key-operated, in a locked cabinet, or provided with protection to prevent use by unauthorized persons. The means will operate a visible indicator and sound a trouble signal whenever the means is in the silence position and there are no telephone circuits in an off hood condition. If a selective talk telephone system is used, such a switch is permitted, provided subsequent telephone circuits going off-hood operate the distinctive off-hook audible signal-sounding appliance.

NFPA 72-1999 3-8.4.1.3.7.6 **Minimum Systems.** As a minimum (for fire service use only), two-way telephone systems are to be common talk (that is, a conference or party line circuit) providing at least one telephone station or jack per floor and at least one telephone station or jack per exit stairway. In buildings equipped with a fire pump(s), a telephone station or jack will be provided in each fire pump room.

NFPA 72-1999 3-8.4.1.3.7.7

Fire Warden Use. If the two-way telephone system is intended to be used by fire wardens in addition to the fire service, the minimum requirement will be a selective talk system (where phones are selected from the fire command center). Systems intended for fire warden use will provide telephone stations or jacks as required for fire service use and additional telephone stations or jacks as necessary to provide at least one telephone station or jack in each voice paging zone. Telephone circuits will be selectable from the fire command center either individually or, if approved by the authority having jurisdiction, by floor or stairwell.

NFPA 72-1999 3-8.4.1.3.7.7

34002 SIGNAL PROCESSING

REFERENCES:

Training Manual on Fire Alarm Systems

Fire Protection Handbook

Fire Alarm Signaling Systems

Manufactures' data

Electrical and Electronic Textbooks

DESCRIPTION:

Understand the processing of alarm signals, which originate in an alarm initiating device and are processed at the control panel to operate the alarm indicating appliance. Understand the basic schematic drawings of noncoded and coded fire alarm systems.

FASSH Chapter 5

Signal Processing

Processing is the dynamic link between signal input and system response. Signal processing is an interpretive function, governing the response of a unit or an entire fire alarm signaling system to the various inputs, which are received by that unit, or system. A control panel will typically receive numerous, varied signal inputs; outputs may vary widely depending on the application of the alarm system.

TYPES OF SIGNALS PROCESSED
Manual Fire Alarm Signals

Manual fire alarm signals in a protective system are typically initiated from manual fire alarm boxes located throughout the protected facility. Signals may also be initiated outside the protective system, for example, from a municipal fire alarm box (as described in NFPA 1221). Processed outputs for manual signals will vary according to the type of protective system.

Some of the differences among manual responses are outlined below. (a) An auxiliary system does not normally use a coded alarm signal. The rationale for this is that auxiliary systems have historically been simple noncoded latching circuits, which did not transmit signals in a serial method to a public fire department. (b) Manual signal initiation is not part of household fire warning equipment. Since most fires occur while the occupants of the household are sleeping, initiation and detection are the only necessary functions of household fire warning equipment. (c) It is assumed in NFPA 72F and NFPA 1221 that other system units or interfaces have processed manual signals prior to retransmission. Noncoded manual alarm inputs are processed in a manner similar to that for automatic alarm inputs.

Automatic Alarm Signals

Automatic alarms are typically initiated by smoke and heat detectors, both of which can be installed and maintained in all the protective signaling systems according to NFPA 72, National Fire Alarm Code. The NFPA signaling system standards refer to special connections for use of integral trouble contacts located on the automatic signal initiating devices. These normally closed contacts should be connected electrically at the end of the initiating circuit to prevent disconnection of initiating devices.

TYPES OF PROCESSING
Guard's Tour Supervisory Service

In guard's tour service, a supervisory service signal is typically initiated through mechanical key operated tour stations. These signals may also be generated electronically or optically for encoding and decoding. "Encoding" can be defined as processing of a signal such as that initiated by a switch contact and subsequent transfer of the signal into information, which can be interpreted by a machine. "Decoding" is the translation of this information back into signals understandable by human beings.

Sprinkler System Waterflow Alarm
and Supervisory Signal Service

In general, a sprinkler system waterflow alarm and supervisory signal service in any protective signaling system must be able to interpret the type of system signal and the particular element, which has operated. Additionally, the supervisory service must identify restoration of the element to the normal position.

Waterflow alarm supervisory signals consist of one or more of the following: gate valve supervision, pressure supervision, water level supervision, water temperature supervision, and pump supervision. The waterflow alarm signals are the only supervisory signals that can be transmitted and processed by auxiliary systems. If additional supervisory services are desired, another type of protective system should be specified.

Trouble Signals

To varying degrees, all systems interpret factors (i.e., "trouble signals") which inhibit alarm transmission or annunciation. Many functions and/ or circuits are supervised against loss of capability, depending on the nature of the system and the complexity of the information processed by that system. In some cases, automatic or manual control of various system functions permits emergency operation under faults that could debilitate the entire system.

34003 SURVEYS FOR FIRE ALARM AND DETECTION SYSTEMS

REFERENCES:

NEMA Guide to Code Reauirements for Fire Protective Signaling and Detection Systems

NFPA 72

NFPA 101

Basic/National Building Code

Fire Alarm Signaling Systems

Standard Building Code

Uniform Building Code

NFPA 72-1999 1-6.1.1

DESCRIPTION:

Know the items required to achieve a complete survey of property for layout of and preparation of plans for an automatic fire detection and alarm system, including type of automatic fire detection devices and alarm indicating appliances that are best suited for the property and in keeping with applicable building and fire codes.

Contact the authority having jurisdiction before a fire alarm system survey.

The authority having jurisdiction is to be notified prior to installation or alteration of equipment or wiring. At its request, complete information regarding the system or system alterations, including specifications, wiring diagrams, battery calculation, and floor plans is to be submitted for approval.

Familiarize yourself with the occupancy classifications of NFPA 101, Life Safety Code, and their protection requirements.

Include an average ambient sound level measurement, on the "A" weighted scale, of all occupied areas of the property to be surveyed.

Understand how to complete an area-by-area analysis according to construction type, ceiling height, flammable material, and ventilation systems.

Understand the different requirements associated with fire alarm systems installed in sprinklered buildings versus non-sprinklered buildings.

Additions to existing buildings will generally be required to meet the requirements for new construction.

NFPA 101-1997 4-1

Classification of Occupancy

NFPA 101-1997 4-1.1

A building or structure will be classified as follows, subject to the ruling of the authority having jurisdiction in case of question as to proper classification in any individual case. Occupancies in special structures will conform to the requirements of the specific occupancy Chapters 8 through 31 except as modified by Chapter 32.

NFPA 101-1997 4-1.2

Assembly. (*For requirements, see Chapters 8 and 9.*) Assembly occupancies include, but are not limited to, all buildings or portions of buildings used for gatherings of 50 or more persons for such purposes as deliberation, worship, entertainment, eating, drinking, amusement, or awaiting transportation. Assembly occupancies also include special amusement buildings regardless of occupant load. (*See 8-4.7 and 9-4.7.*)

Assembly occupancies include the following:

Armories	Libraries
Assembly Halls	Mortuary chapels
Auditoriums	Motion picture theaters
Bowling lanes	Museums
Club rooms	Passenger stations and terminals of air, surface, underground, and marine public transportation facilities
College and university classrooms 50 persons and over	Places of religious worship
Conference rooms	Pool rooms
Courtrooms	Recreation piers
Dance halls	Restaurants
Drinking establishments	Skating rinks
Exhibition halls	Theaters
Gymnasiums	

Occupancy of any room or space for assembly purposes by fewer than 50 persons in a building or other occupancy and incidental to such other occupancy will be classified as part of the other occupancy and will be subject to the provisions applicable thereto.

NFPA 101-1997 4-1.3

Educational. (*For requirements, see Chapters 10 and 11.*) Educational occupancies include all buildings or portions of buildings used for educational purposes through the twelfth grade by six or more persons for four or more hours per day or more than 12 hours per week.

Educational occupancies include the following:

Academies	Nursery Schools
Kindergartens	Schools

Other occupancies associated with educational institutions will be in accordance with the appropriate parts of this *Code.*
In cases where instruction is incidental to some other occupancy, the section of this *Code* governing such other occupancy will apply.

NFPA 101-1997 4-1.4

Health Care. (*For requirements, see Chapters 12 and 13.*) Health care occupancies are those used for purposes such as medical or other treatment or care of persons suffering from physical or mental illness, disease, or infirmity; and for the care of infants, convalescents, or infirm aged persons. Health care occupancies provide sleeping facilities for four or more occupants and are occupied by persons who are mostly incapable of self-preservation because of age, physical or mental disability, or because of security measures not under the occupants' control.

Health care occupancies include the following:

Hospitals
Limited care facilities
Nursing homes

Health care occupancies also include ambulatory health care centers. (*See Sections 12-6 and 13-6.*)

NFPA 101-1997 4-1.5

Detention and Correction. (*For requirements, see Chapters 14 and 15.*) Detention and correctional occupancies are used to house individuals under varied degrees of restraint or security and are occupied by persons who are mostly incapable of self-preservation because of security measures not under the occupants' control. Detention and correctional occupancies include the following:

Adult and juvenile substance abuse centers
Adult local detection facilities
Adult and juvenile work camps

Juvenile community residential centers
Adult community residential centers
Juvenile detention facilities
Adult correctional institutions
Juvenile training schools
Chapters 14 and 15 address the residential housing areas of the detention and correctional occupancy as defined by 14-1.3 and 15-1.3. Other uses within detention and correctional facilities, such as gymnasiums or industries, will be in accordance with the appropriate chapter of the *Code. (See 14-1.2.1 and 15-1.2.1.)*

NFPA 101-1997 4-1.6

Residential. (*For requirements, see Chapters 16 through 23.*) Residential occupancies are those occupancies in which sleeping accommodations are provided for normal residential purposes and include all buildings designed to provide sleeping accommodations.

Exception: Those classified under health care or detention and correctional occupancies.

Residential occupancies are treated separately in this *Code* in the following groups:

Hotels, motels, and dormitories (*Chapters 16 and 17*);
Apartment buildings (*Chapters 18 and 19*);
Lodging or rooming houses (*Chapter 20*);
One- and two-family dwellings (*Chapter 21*); and
Board and care facilities (*Chapters 22 and 23*).

NFPA 101-1997 4-1.7

Mercantile (*For requirements, see Chapters 24 and 25.*) Mercantile occupancies include stores, markets, and other rooms, buildings, or structures for the display and sale of merchandise.
Mercantile occupancies include the following:

Auction rooms
Shopping centers
Department stores
Supermarkets
Drugstores

NFPA 101-1997 4-1.8

Business (*For requirements, see Chapters 26 and 27.*) Business occupancies are those used for the transaction of business (other than those covered under Mercantile), for the keeping of accounts and records, and for similar purposes.

Business occupancies include the following:

Air traffic control towers (ACTCs)

Courthouses

City halls

Dentists' offices

College and university instructional Buildings, classrooms under

50 persons, and instructional laboratories

Doctors' offices

General offices

Outpatient clinics, ambulatory

Town halls.

NFPA 101-1997 4-1.9

Industrial. (*For requirements, see Chapter 28.*) Industrial occupancies include factories making products of all kinds and properties devoted to operations such as processing, assembling, mixing, packaging, finishing or decorating, and repairing.
Industrial occupancies include the following:

Dry cleaning plants	Laundries
Factories of all kinds	Power plants
Food processing plants	Pumping stations
Gas plants	Refineries
Hangars (for servicing/maintenance)	Sawmills
	Telephone exchanges

NFPA 101-1997 4-1.10

Storage. (*For requirements, see Chapter 29.*) Storage occupancies include all buildings or structures utilized primarily for the storage or sheltering of goods, merchandise, products, vehicles, or animals.
Storage occupancies include the following:

Barns	Hangars (for storage only)
Bulk oil storage	Parking structures
Cold storage	Stables
Freight terminals	Truck and marine terminals
Grain elevators	Warehouses

NFPA 101-1997 4-1.11

Day-Care. *(For requirements, see Chapters 30 and 31.)* Day-care occupancies include all buildings or portions of buildings in which four or more clients receive care, maintenance, and supervision, by other than their relatives or legal guardians, for less than 24 hours per day. Day-care occupancies include the following:

Child day-care occupancies
Kindergarten classes that are incidental to a child day-care
Adult day-care occupancies, except where occupancy is part of a health care occupancy
Nursery schools
Day-care homes

In cases where care is incidental to some other occupancy, the section of this *Code* governing such other occupancy will apply.

NFPA 101-1997 4-1.12

Mixed Occupancies. Where two or more classes of occupancy occur in the same building or structure and are intermingled so that separate safeguards are impractical, means of egress facilities, construction, protection, and other safeguards will comply with the most restrictive life safety requirements of the occupancies involved.

Exception: An occupancy incidental to operations in another occupancy will be permitted to be considered as part of the predominant occupancy and will be subject to the provisions of this Code that apply to the predominant occupancy.

NFPA 101-1997 4-2.2

Classification of Hazard of Contents.

NFPA 101-1997 4-2.2.1

The hazard of contents of any building or structure will be classified as low, ordinary, or high in accordance with 4-2.2.2, 4-2.2.4, and 4-2.2.4.

NFPA 101-1997 4-2.2.2

Low Hazard. Low hazard contents will be classified as those of such low combustibility that no self-propagating fire therein can occur.

NFPA 101-1997 4-2.2.3

Ordinary Hazard. Ordinary hazard contents will be classified as those that are likely to burn with moderated rapidity or to give off a considerable volume of smoke.

NFPA 101-1997 4-2.2.4

High Hazard. High hazard contents will be classified as those that are likely to burn with extreme rapidity or from which explosions are likely. *(For means of egress requirements, see Section 5-11.)*

NFPA 101-1997 7-6.2.8

Complete Smoke Detection Systems. Where a complete smoke detection system is required by another section of this *Code*, automatic detection of smoke in accordance with NFPA 72, *National Fire Alarm Code*, will be provided in all occupiable areas, common areas, and work spaces in those environments suitable for proper smoke detector operation.

NFPA 101-1997 7-6.3.3

Where a standard evacuation signal is required by another section of this *Code*, the evacuation signal will be the standard fire alarm evacuation signal described in NFPA *72 National Fire Alarm Code.*

NFPA 101-1997 7-6.3.6

Audible alarm notification appliances will be of such character and so distributed as to be effectively heard above the average ambient sound level occurring under normal conditions of occupancy.

NFPA 101-1997 7-6.3.7

Audible alarm notification appliances will produce signals that are distinctive from audible signals used for other purposes in the same building.

Notes

34004 FIRE ALARM SYSTEM MAINTENANCE

REFERENCES:

NFPA 72

Electrical Textbooks

DESCRIPTION:

Maintain a fire alarm system. Know the tests and inspections, which will identify components that have become undependable or inoperative. Know which components are field serviceable, methods of cleaning, checking, operation and adjustment of the system and its components to keep the system in an operative condition.

NFPA 72-1999 7-1.1.1

Inspection, testing, and maintenance programs will satisfy the requirements of NFPA 72-1999, conform to the equipment manufacturer's recommendations, and verify proper operation of the fire alarm system.

NFPA 72-1999 7-1.2

The owner or owner's designated representative is responsible for inspection, testing, and maintenance of the system and alterations or additions to this system. The delegation of responsibility is to be in writing, with a copy of such delegation provided to the authority having jurisdiction upon request.

NFPA 72-1999 7-1.6.1

Initial Acceptance Testing. All new systems are to be inspected and tested in accordance with the requirements of NFPA 72-1999 chapter 7.

Familiarize yourself with the test methods described in chapter 7 of NFPA 72 such as:

NFPA 72-1999 Table 7-2.2

14. Alarm Notification Appliances

a. Audible

Measure sound pressure level with sound level meter meeting ANSI S-1.4a, *Specifications for Sound Level Meters*, Type 2 requirements. Levels throughout the protected area are to be measured and recorded.

b. Audible Textual

Sound pressure level is to be measured with sound level meter meeting ANSI S-1.4a, *Specifications for Sound Level*

Meters, Type 2 requirements. Levels throughout the protected area are to be measured and recorded.

Audible information will be verified to be distinguishable and understandable.

c. Visible

Test is to be in accordance with manufacturer's instructions. Device locations are to be verified to be per approved layout and it is to be confirmed that no floor plan changes affect the approved layout.

NFPA 72-1999 7-3.2

NFPA 72-1999 A-7-3.2

Testing. Testing is to be performed in accordance with the schedules in NFPA 72-1999 Chapter 7 or more often if required by the authority having jurisdiction. If a remotely monitored fire alarm control unit specifically listed for the application performs automatic testing at least weekly, the manual testing frequency will be permitted to be extended to annual. NFPA 72-1999 Table 7-3.2 applies, except that devices or equipment that are inaccessible for safety considerations, (for example, continuous process operations, energized electrical equipment, radiation, excessive height) are to be tested during scheduled shutdowns if approved by the authority having jurisdiction but not more than every 18 months.

It is suggested that the annual test be conducted to segments such that all devices are tested annually.

NFPA 72-1999 7-3.2.1

Detector sensitivity is to be checked within 1 year after installation and every alternate year thereafter. After the second required calibration test, where the sensitivity test indicate that the detector has remained within its listed and marked sensitivity range (or 4 percent obscuration light gray smoke, if not marked), the length of time between calibration tests will be permitted to be extended to a maximum of 5 years. Where the frequency is extended, records of detector-caused nuisance alarms and subsequent trends of these alarms will be maintained. In zones or in areas where nuisance alarms show any increase over the previous year, calibration tests are to be performed.

To ensure that each smoke detector is within its listed and marked sensitivity range, it is to be tested using any of the following methods:

(a) A calibrated test method; or

(b) The manufacturer's calibrated sensitivity test instrument; or

(c) Listed control equipment arranged for the purpose; or

(d) A smoke detector/control unit arrangement whereby the detector causes a signal at the control unit where its sensitivity is outside its acceptable sensitivity range; or

(e) Other calibrated sensitivity test method acceptable to the authority having jurisdiction.

Detectors found to have a sensitivity reading outside the listed and marked sensitivity range are to be cleaned and recalibrated or replaced except detectors listed as field adjustable are permitted to be either adjusted within the listed and marked sensitivity range and cleaned and recalibrated, or replaced. This requirement does not apply to single station detectors referenced in NFPA 72-1999 7-3.3 (*single-station smoke detectors installed in one- and two- family dwelling units*) and Table 7-2.2 (**Test Methods**).

Detector sensitivity will not be tested or measured using any device that administers an unmeasured concentration of smoke or other aerosol into the detector.

NFPA 72-1999 7-4.1

Fire alarm system equipment is to be maintained in accordance with the manufacturer's instructions. The frequency of maintenance depends on the type of equipment and local ambient conditions.

NFPA 72-1999 7-4.3

All apparatus requiring rewinding or resetting to maintain normal operation is to be restored to normal as promptly as possible after each test and alarm. All test signals received are to be recorded to indicate date, time, and type.

NFPA 72-1999 7-4.4

The retransmission means as defined in NFPA 72-1999 Section 5-2 (**Fire Alarm Systems for Central Station Service**) is to be tested at intervals of not more than 12 hours. The retransmission signal and the time and date of the retransmission is to be recorded in the central station except; where the retransmission means is the public switched telephone network, it is permitted to be tested weekly to confirm its operation to each public fire service communications center.

NFPA 72-1999 7-4.4

34005 FIRE ALARM SYSTEM WIRING

<table>
<tr><td>REFERENECES:</td><td>DESCRIPTION:</td></tr>
<tr><td>NFPA 70</td><td>Understand the proper types of wire, cable or conduit used on a fire alarm system and where, in compliance with the codes, each is permitted and the correct and incorrect method of field wiring system components. Prepare an electrical riser and plan view diagram of a multi-zone, multi-story fire alarm system consisting of manual fire alarm boxes, heat detectors, four wire smoke detectors and a variety of audible and visible appliances. The diagrams are to show the number and size of conductors and conduit and the location of each system component.</td></tr>
</table>

NFPA 70-1999 760-2

Fire Alarm Circuit: The portion of the wiring system between the load side of the overcurrent device or the power-limited supply and the connected equipment of all circuits powered and controlled by the fire alarm system. Fire alarm circuits are classified as either nonpower-limited or power-limited.

NFPA 70-1999 760-10

Fire Alarm Circuit Identification: Fire alarm circuits will be identified at terminal and junction locations, in a manner that will prevent unintentional interference with the signaling circuit during testing and servicing.

NFPA 70-1999 760-2

Fire Alarm Circuit Integrity (CI) Cable. Cable used in fire alarm systems to ensure continued operation of critical circuits during a specified time under fire conditions.

NFPA 70-1999 760-30

Multiconductor NPLFA Cables. Multiconductor nonpower-limited fire alarm cables that meet the requirements of Section 760-31 will be permitted to be used on fire alarm circuits operating at 150 volts or less and will be installed in accordance with NFPA 70-1999 760-30 (a) and (b).

NFPA 70-1999 760-31 (f)

Fire Alarm Circuit Integrity (CI) Cable. Cables suitable for use in fire alarm systems to ensure survivability of critical circuits during a specified time under fire conditions will be listed as circuit integrity (CI) cable. Cables identified in NFPA 70-1999 760-31 (c), (d), and (e) meeting the requirements for circuit integrity will have the additional classification using the suffix "CI" (for example, NPLFP-CI, NPLFR-CI, and NPLF-CI).

NOTE 1: This cable may be used for fire alarm circuits to comply with the survivability requirements of NFPA 72-1996 *National Fire Alarm Code*, subsections 3-2.4, 3-4.4, 3-12.4, and 3-12.4.3, that the cable maintain its electrical function during fire conditions for a period of time.

NOTE 2: One method of defining circuit integrity (CI) cable is by establishing a minimum 2-hour fire resistance rating for the cable when tested in accordance with UL 2196-1995 *Standard for Tests of Fire Resistive Cable*.

NFPA 70-1999 760-31 (g)

NPLFA Cable Markings. Multiconductor nonpower-limited fire alarm cables will be marked in accordance with NFPA 70-1999 Table 760-31(g). Nonpower-limited fire alarm circuit cables will be permitted to be marked with a maximum usage voltage rating of 150 volts. Cables that are listed for circuit integrity will be identified with the suffix "CI" as defined in NFPA 70-1999 760-31 (f).

*NFPA 70-1999
Table 760-31 (g)*

NPLFA Cable Markings.

Cable Marking	Type	Reference
NPLFA	Nonpower-limited fire alarm Circuit cable for use in other space used for environmental air	Sections 760-31(d) and (g).
NPLFR	Nonpower-limited fire alarm circuit riser cable	Sections 760-31(e) and (g).
NPLF	Nonpower-limited fire alarm circuit cable	Sections 760-31(f) and (g).

NOTE: Cables identified in (c), (d), and (e) meeting the requirements for circuit integrity shall have the additional classification using suffix "CI" (for example, NPLFP-CI, NPLFR-CI, and NPLF-CI).

NOTE: Cable types are listed in descending order of fire-resistance rating.

NFPA 70-1999 760-54

Installation of Conductors and Equipment.

(a) Separation from Electric Light, Power, Class 1, and NPLFA, and Medium Power Network-Powered Broadband Communications Circuit Conductors.

(1) In Cables, Compartments, Enclosures, Outlet Boxes, or Raceways.

Power-limited circuit conductors will not be placed in any cable, compartment, enclosure, outlet box, raceway, or similar fitting with conductors of electric light, power, Class 1, nonpower-limited fire alarm circuit conductors, or medium power network-powered broadband communications circuits;

Except where the conductors of the electric light, power, Class 2, nonpower-limited fire alarm circuit conductors, and medum power network-powered broadband communications circuit conductors are separated by a barrier from the power-limited fire alarm circuits. In enclosures, power-limited fire alarm circuits will be permitted to be installed in a raceway within the enclosure to separate them from Class1, electric light, power, nonpower-limited fire alarm circuits, and medum power network-powered broadband communications circuit conductors; and;

Except conductors in compartments, enclosures, device boxes, outlet boxes, or similar fittings, where electric light, power, Class 1, nonpower-limited fire alarm circuit conductors, or medum power network-powered broadband communications circuit conductors are introduced solely to connect to the equipment connected to power-limited circuits to which the other conductors are connected and

a. The electric light, power, Class 1, and nonpower-limited fire alarm circuit conductors are routed to maintain a minimum of 0.25 in. (6.35 mm) separation from the conductors and cables of power-limited fire alarm circuits, or

b. The circuit conductors operate at 150 volts or less to ground and also comply with one of the following:

1. The fire alarm power-limited circuits are installed using Types FPL, FPLR, FPLP, or permitted substitute cables, provided these power-limited cable conductors extending beyond the jacket are separated by a minimum of 0.25 in. (6.35 mm) or by a nonconductive sleeve or nonconductive barrier from all other conductors, or

2. The fire alarm power-limited circuit conductors are installed as nonpower-limited fire alarm circuits in accordance with NFPA 70-1999 760-25; and

Except conductors entering compartments, enclosures, device boxes, outlet boxes, or similar fittings, where electric light, power, Class 1, nonpower-limited fire alarm circuit conductors, or medum power network-powered broadband communications circuit conductors are introduced solely to connect to equipment connected to power-limited fire alarm circuits or to other circuits controlled by the fire alarm system to which the other conductors in the enclosure are connected. If the conductors must enter an enclosure that is provided with a single opening, they will be permitted to enter through a single fitting (such as a tee) provided the conductors are separated from the conductors of the other circuits by a continuous and firmly fixed nonconductor, such as flexible tubing.

NFPA 70-1999
Table 760-51

Cable Uses and Permitted Substitutions

Cable Type	Use	References	Permitted Substitutions
FPLP	Power-limited fire alarm plenum cable	760-61 (a)	MPP, CMP, CL3P
FPLR	Power-limited fire alarm riser cable	760-61 (b)	MPP, CMP, FPLP, CL3P, MPR, CMR, CL3R
FPL	Power-limited fire alarm cable	760-61(c)	MPP, CMP, FPLP, CL3P, MPR, CMR, CL3R, FPLR, MPG, MP, CMG, CM, PLTC, CL3

34006 EMERGENCY EVACUATION SIGNALS

REFERENCES: *NFPA 72* *ANSI S1.4a 1985*	**DESCRIPTION:** Understand the recommended fire alarm evacuation signal, its purpose, where it is used, the recommended sound level in various occupancies and modes and how to determine the sound level needed.

NFPA 72-1999 1-4

NFPA 72-1999 A-1-4

Evacuation. The withdrawal of occupants from a building. Evacuation does not include the relocation of occupants within a building.

NFPA 72-1999 1-4

Evacuation Signal. Distinctive signal intended to be recognized by the occupants as requiring evacuation of the building.

NFPA 72-1999 1-5.4.7

NFPA 72-1999 A-1-4.7(b)

Distinctive Signals. Audible alarm notification appliances for a fire alarm system will produce signals that are distinctive from other similar appliances used for other purposes in the same area. The distinction among signals will be as follows:

(a) Fire alarm signals will be distinctive in sound from other signals. Their sound will not be used for any other purpose. The requirements of NFPA 72-1999 3-8.4.1.2.1 (ANSI S3.41, *Audible Emergency Evacuation Signal*),

(b) Supervisory signals will be distinctive in sound from other signals. Their sound is not to be used for any other purpose, except that a supervisory signal sound is permitted to be used to indicate a trouble condition. Where the same sound is used for both supervisory signals and trouble signals, the distinction between signals is to be by other appropriate means such as visible annunciation, and/or

(c) Fire alarm, supervisory, and trouble signals are to take precedence, in that respective order of priority, over all other signals except that signals from hold-up alarms or other life threatening signals are permitted to take precedence over supervisory and trouble signals if acceptable to the authority having jurisdiction.

NFPA 72-1999 3-8.4.1.2

NFPA 72-1999 3-8.4.1.2.1

Distinctive Evacuation Signal. To meet the requirements of NFPA 72-1999 1-5.4.7 (**Distinctive Signals**), the fire alarm signal used to notify building occupants of the need to evacuate (leave the building) will be in accordance with ANSI S3.41, *Audible Emergency Evacuation Signal.*

NFPA 72-1999 3-8.4.1.2.2

The use of the American National Standard Audible emergency Evacuation Signal will be restricted to situations where it is desired that all occupants hearing the signal evacuate the building immediately. It will not be used where, with the approval of the authority having jurisdiction, the planned action during a fire emergency is not evacuation, but is the relocation of occupants or their protection in place as directed by the firefighting personnel.

NFPA 72-1999 A-3-4.1.2

NFPA 72-1999
Figures A-3-6.4.1.2(a)(b)(c)

The use of the distinctive three-pulse temporal pattern fire alarm evacuation signal required by NFPA 72-1999 3-8.4.1.2.2 became effective July 1, 1996, for new systems installed after that date. It had previously been recommended for this purpose by this code since 1979. It has since been adopted as both an American National Standard (ANSI S3.41, *Audible Emergency Evacuation* Signal) and an International Standard (ISO 8201, *Audible Emergency Evacuation Signal*).

Copies of both of these standards are available from the Standards Secretariat, Acoustical Society of American, 335 East 45th Street, New York, NY 10017-3483. Telephone 212-661-9404 ext. 562.

The standard fire alarm evacuation signal is a three-pulse temporal pattern using any appropriate sound. The pattern consists of an "on" phase (a) lasting 0.5 second ＂ 10 percent followed by an "off" phase (b) lasting 0.5 second ＂ 10 percent, for three successive "on" periods, which are then followed by an "off" phase (c) lasting 1.5 second ＂ 10 percent [*see NFPA 72-1999 Figures A-3-8.4.1.2 (a) and A-3-8.4.1.2 (b)*]. The signal should be repeated for a period appropriate for the purposes of evacuation of the building, but for not less than 180 seconds. A single-stroke bell or chime sounded at "on" intervals lasting 1 second ＂ 10 percent, with a 2-second ＂ 10 percent "off" interval after each third "on" stroke, may be permitted [*see NFPA 72-1999 Figure A-3-8.4.1.2(c)*].

The minimum repetition time is permitted to be manually interrupted.

NFPA 72-1999 3-8.4.1.3.5.6.4 **Loudspeakers.** Each enclosed stairway exceeding two stories in height is to be equipped with loudspeakers connected to a separate paging zone.

NFPA 72-1999 3-8.4.1.3.6

NFPA 72-1999 3-8.4.1.3.6.1

NFPA 72-1999 A-3-8.4.1.3.6.1 **Evacuation Signal Zoning.** Undivided fire or smoke areas will not be divided into multiple evacuation signaling zones. This does not prohibit the provision of multiple notification appliance circuits within an evacuation zone.

NFPA 72-1999 3-8.4.1.3.6.2 If multiple notification appliance circuits are provided within a single evacuation signaling zone, all of the notification appliances within the zone are to be arranged to activate simultaneously, either automatically or by actuation of a common, manual control except where the different notification appliance circuits within an evacuation signaling zone perform separate functions, for example, presignal and general alarm signals, predischarge and discharge signals.

NFPA 72-1999 4-3.1.1 An average ambient sound pressure level greater than 105dBA requires the use of a visible signal appliance(s) in accordance with NFPA 72-1999 Section 4-4 (**Visible Characteristics, Public Mode**).

NFPA 72-1999 4-3.2

NFPA 72-1999 A-6-3.2 **Public Mode Audible Requirements.** The typical average ambient sound level for the occupancies specified in NFPA 72-1999 Table A-4-3.2 (*Average Ambient Sound Level According to Location*) are intended only for design guidance purposes.

The typical average ambient sound levels specified should not be used in lieu of actual sound level measurements.

NFPA 72-1999 4-3.2.2 To ensure that audible public mode signals are clearly heard, they will have a sound level at least 15dBA above average ambient sound level or 5 dBA above the maximum sound level having a duration of at least 60 seconds, whichever is greater, measured 5 ft (1.5 m) above the floor in the occupiable area, except audible alarm notification appliances installed in elevator cars are permitted to use the audibility criteria for private mode appliances found in NFPA 72-1999 4-3.3.2 (**Private Mode**). If approved by the AHJ, audible alarm notification appliances installed in restrooms are permitted to use the audibility criteria for private mode appliances detailed in NFPA 72-1999 4-3.3.2

(Private Mode). If permitted by the AHJ, a fire alarm system arranged to stop or reduce ambient noise are permitted to produce a sound level at least 15dBA above the reduced average ambient sound level or 5dBA above the maximum sound level having a duration of at least 60 seconds after reduction of the ambient noise level, whichever is greater, measured 5 feet above the floor in the occupiable area. Visible notification appliances will be installed in the affected areas in accordance with NFPA 72-1999 Sections 4-4 **(Visible Characteristics, Public Mode)** or 4-5 **(Visible Characteristics, Private Mode)**. Relays, circuits, or interfaces necessary to stop or reduce ambient noise will meet the requirements of NFPA 72-1999 Chapter 1 **(Fundamentals of Fire Alarm Systems)** and Chapter 3 **(Protected Premises Fire Alarm Systems)**.

34007 COMBINATION SYSTEMS

REFERENCES:

NFPA 72

Fire Protection Handbook

*Fire Alarm
Signaling Systems*

DESCRIPTION:

Understand the combination systems that are permitted by the codes such as paging, music, intruder alarm, energy management and process monitoring; their purpose, restrictions, basic requirements, supervision and priority of signals.

NFPA 72-199 1-4

NFPA 72-1999 A-1-4

Combination System. A fire alarm system whose components might be used, in whole or in part, in common with a nonfire signaling system, such as a security, card access, closed circuit television, sound reinforcement, background music, paging, sound masking, building automation, and time and attendance.

NFPA 72-1999 3-8.2

Combination Systems.

NFPA 72-1999 3-8.2.1

NFPA 72-1999 A-3-2.1

Fire alarm systems are permitted to share components, equipment, circuitry, and installation wiring with non-fire alarm systems. The provisions of this paragraph apply to the types of equipment used in common for fire alarm, sprinkler supervisory, or guard's tour service and for other systems such as intrusion alarm or coded paging systems, and to methods of circuit wiring common to both types of systems.

NFPA 72-1999 3-8.2.2

Where common wiring is employed for combination systems, the equipment for other than fire alarm systems is permitted to be connected to the common wiring of the system. Short circuits, open circuits, or grounds in this equipment or between this equipment and the fire alarm system wiring is not allowed to interfere with the monitoring for integrity of the fire alarm system or prevent alarm, supervisory, or fire safety control signal transmissions.

NFPA 72-1999 3-8.2.3

To maintain the integrity of fire alarm system functions, the provision for removal, replacement, failure, or maintenance procedure on any supplementary hardware, software, or circuit(s) is not allowed to impair the required operation of the fire alarm systems, except where hardware, software, or circuit(s) is listed for fire alarm use.

NFPA 72-1999 3-8.2.4

Speakers used as alarm notification appliances on fire alarm systems are not to be used for nonemergency purposes, except where the fire command center is constantly attended by a trained operator, selective paging is permitted as approved by the AHJ and except where the following conditions are met:

- The speakers and associated audio equipment are installed or located with safeguards to prevent tampering or misadjustment of those components essential to intended operation for fire.
- The monitoring integrity requirements of NFPA 72-1999 1-5.8 (**Monitoring Integrity of Installation Conductors and Other Signaling Channels**) and 3-8.4.1.3.2 (**Distinctive Evacuation Signal**) will continue to be met while the system is used for nonemergency purposes.
- It is permitted by the AHJ.

NFPA 72-1999 3-8.2.5

In combination systems, fire alarm signals are to be distinctive, clearly recognizable, and take precedence over any other signal even when a non-fire alarm signal in initiated first.

Table A-4-3.2 Average Ambient Sound Level According to Location

Location	Average Ambient Sound Level (dBA)
Business occupancies	55
Education occupancies	45
Industrial occupancies	80
Institutional occupancies	50
Mercantile occupancies	40
Piers and water-surrounded structures	40
Places of assembly	55
Residential occupancies	35
Storage occupancies	30
Thoroughfares, high density urban	70
Thoroughfares, medium density urban	55
Thoroughfares, rural and suburban	40
Tower occupancies	35
Underground structures and windowless buildings	40
Vehicles and vessels	50

Table A-2-8.3 Recommended Coded Signal Designations

Location	Coded Signal
Fourth floor	2-4
Third floor	2-3
Second floor	2-2
First floor	2-1
Basement	3-1
Sub-basement	3-2

Table A-2-4.1 Spectrum Wavelength Ranges

Radiant Energy	μm
Ultraviolet	0.1 – 0.35
Visible	0.36 – 0.75
Infrared	0.76 – 220

Fire Alarm Systems Glossary

Accessible Area of Refuge. An area of refuge that complies with the accessible route requirements of CABO/ANSI A117.1, *American National Standard for Accessible and Usable Buildings and Facilities.*

Accessible Means of Egress. A path of travel, usable by a person with a severe mobility impairment, that leads to a public way or an area of refuge.

Acknowledge. To confirm that a message or signal has been received such as by the pressing of a button or the selection of a software command.

Active Multiplex System. A multiplexing system in which signaling devices such as transponders are employed to transmit status signals of each initiating device or initiating device circuit within a prescribed time interval so that lack of receipt of such signal may be interpreted as a trouble signal.

Addition. An extension or increase in floor area or height of a building or structure.

Addressable Device. A fire alarm system component with discrete identification that can have its status individually identified or that is used to individually control other functions.

Addressable Control Device. A signaling system output device which, when operating with a compatible control unit, is used to control individual preselected electrical circuits such as audible or visual alarm signaling appliances, fan circuits, or door release circuits.

Addressable Intelligent Device. A signaling system data input device which, when communicating with a compatible control unit, may have its status individually identified by the control unit.

Addressable System. A system, which uses a signaling technique that, allows a control unit to identify a specific initiating device or group of devices by location.

Addressable System Smoke Detector. A smoke detector, which, in addition to providing alarm and trouble outputs to a control unit, may be uniquely identified by location to the main control unit.

Adverse Condition. Any condition occurring in a communications or transmission channel that interferes with the proper transmission or interpretation, or both, of status change signals at the supervising station. See also *Trouble Signal.*

Agent Release. The release of a chemical substance (i.e., Halon, FM200) capable of producing a chemical reaction which will extinguish a fire. Agent release is most often used in areas where water release would cause damage to electrical equipment.

Agglomerate. To gather into a ball, mass, or cluster.

Air Sampling-Type Detector. A detector that consists of a piping or tubing distribution network that runs from the detector to the area(s) to be protected. An aspiration fan in the detector housing draws air from the protected area back to the detector through air sampling ports, piping, or tubing. At the detector, the air is analyzed for fire products.

Alarm. A warning of fire danger.

Alarm Initiating Device. A device which when actuated, initiates an alarm. Such devices, dependant upon their type, may be operated manually or automatically. Automatic initiating devices may respond to smoke, heat, or waterflow, for example.

Alarm Service. Service required following the receipt of an alarm.

Alarm Signal. A signal, indicating an emergency requiring immediate action, such as a signal indicative of fire.

Alarm System. A combination of compatible initiating devices, control panels, and indicating appliances, designed and installed to produce an alarm signal in the event of fire.

Alarm Verification Feature. A feature of automatic fire detection and alarm systems to reduce unwanted alarms wherein smoke detectors report alarm conditions for a minimum period of time, or confirm alarm conditions within a given time period after being reset, in order to be accepted as a valid alarm initiation signal.

Alert Tone. An attention-getting signal to alert occupants of the pending transmission of a voice message.

All-Call Mode. A selector switch on the fire alarm control unit which allows all speaker circuits to be used for an "All-Call Page" transmission.

Analog Initiating Device (Sensor). An initiating device that transmits a signal indicting varying degrees of conditions as contrasted with a conventional initiating device, which can only indicate an on/off condition.

Analog Intelligent Control Panel. An addressable, programmable fire alarm control unit which determines when and whether a device or system is in alarm.

Analog System. A fire alarm system which measures how much of a substance exists at an automatic initiating device contrasted with a conventional system which can only determine whether the initiating device is on or off (in alarm or not in alarm). Measurements may include smoke density (percent obscuration), temperature, water level, air pressure, etc.

Analog Smoke Detector. A system smoke detector capable of communicating information regarding measured smoke level to a control unit. This type of detector is capable of sending signals to the control unit, which indicate the analog level of smoke within the detector. An analog smoke detector is typically used as part of an addressable system but differs from the Addressable Smoke Detector in that it is capable of communicating the level of smoke as well as its discrete address and its alarm and/or trouble condition.

Annunciator. A unit containing one or more indicator lamps, alphanumeric displays, or other equivalent means in which indication provides status information about a circuit, condition, or location.

ANSI. American National Standards Institute, 1430 Broadway, New York, NY 10018.

Approved. Acceptable to the authority having jurisdiction. The National Fire Protection Association does not approve, inspect, or certify any installations, procedures, equipment, or materials; nor does it approve or evaluate testing laboratories. In determining the acceptability of installations, procedures, equipment, or materials, the authority having jurisdiction may base acceptance on compliance with NFPA or other appropriate standards. In the absence of such standards, said authority may require evidence of proper installation, procedure, or use. The authority having jurisdiction of an organization concerned with product evaluations that are in a position to determine compliance with appropriate standards for the current production of listed items.

Assembly Occupancy. Assembly occupancies include, but are not limited to, all buildings or portions of buildings used for gatherings of 50 or more persons for such purposes as deliberation, worship, entertainment, eating, drinking, amusement, or awaiting transportation. Assembly occupancies also include special amusement buildings regardless of occupant load. Assembly occupancies include armories, libraries, assembly halls, mortuary chapels, auditoriums, motion picture theaters, bowling lanes, museums, club rooms, passenger stations and terminals for air, surface, underground, and marine public transportation facilities, college and university classrooms, 50 persons and over, conference rooms, places of religious worship, courtrooms, pool rooms, dance halls, recreation piers, drinking establishments, restaurants, exhibition halls, skating rinks, gymnasiums, and theaters.

ASTM. American Society for Testing and Materials, 1916 Race Street, Philadelphia, PA 19103.

Atrium. A floor opening or series of floor openings connecting two or more stories that is covered at the top of the series of openings and is used for purposes other than an enclosed stairway; elevator hoistway; escalator opening; or utility shaft used for plumbing, electrical, air conditioning, or communication facilities.

Audible Signal. The sound made by one or more audible notification appliances such as bells, chimes, horns, and speakers, which respond to the operation of an initiating device.

Audio Voice Link (AVL). Allows the fire alarm control unit to play a single channel of different voice messages from a library of words.

Authority Having Jurisdiction. The organization, office, or individual responsible for approving equipment, an installation, or a procedure. The phrase "authority having jurisdiction" is used in NFPA documents in a broad manner, since jurisdictions and approval agencies vary, as do their responsibilities. Where public safety is primary, the authority having jurisdiction may be a federal, state, local, or other regional department, or individual such as a fire chief; fire marshal; chief of a fire prevention bureau, labor department, or health department; building official; electrical inspector; or others having statutory authority. For insurance purposes, an insurance inspection department, rating bureau, or other insurance company representative may be the authority having jurisdiction, In many circumstances, the property owner or his or her designated agent assumes the role of the authority having jurisdiction; at government installations, the commanding officer or departmental official may be the authority having jurisdiction.

Automatic. Providing a function without the necessity of human intervention.

Automatic Extinguishing System Supervisory Device. A device that responds to abnormal conditions that could affect the proper operation of an automatic sprinkler system or other fire extinguishing system(s) or suppression system(s) including, but not limited to, control valves; pressure levels; liquid agent levels and temperatures; pump power and running; engine temperature and overspeed; and room temperature.

Automatic Extinguishing or Suppression System Operation Detector. A device that automatically detects the operation of a fire extinguishing or suppression system by means appropriate to the system employed.

Automatic Fire Detector. A device designed to detect the presence of a fire signature and to initiate action. For the purpose of this code, automatic fire detectors are classified as follows:

> **Automatic Extinguishing or Suppression System Operation Detector.** A device that automatically detects the operation of a fire extinguishing or suppression system by means appropriate to the system employed.
>
> **Fire-Gas Detector.** A device that detects gases produced by a fire.
>
> **Heat Detector.** A fire detector that senses heat produced by burning substances. Heat is the energy produced by combustion that causes substances to rise in temperature.
>
> **Other Fire Detectors.** Devices that detect a phenomenon other than heat, smoke, flame, or gases produced by a flame.
>
> **Radiant Energy-Sensing Fire Detector.** A device that detects radiant energy (such as ultraviolet, visible, or infrared) that is emitted as a product of combustion reaction and obeys the laws of optics.
>
> **Smoke Detector.** A device that detects visible or invisible particles of combustion.

Automatic Fire Alarm System. A system in which all or some of the circuits are actuated by automatic devices, such as fire detectors, smoke detectors, heat detectors, and flame detectors.

Automatic Float Charger. Provides secondary (standby) power for temporary operation of the control unit following the loss of primary (main) power. The float charger keeps the secondary (standby) batteries charged during normal operation.

Auxiliary Box. A fire alarm box that can be operated from one or more remote actuating devices.

Auxiliary Fire Alarm System. A system connected to a municipal fire alarm system for transmitting an alarm of fire to the public fire service communications center. Fire alarms from an auxiliary fire alarm system are received at the public fire service communications center on the same equipment and by the same methods as alarms transmitted manually from municipal fire alarm boxes located on streets.

Local Energy Type. An auxiliary system that employs a locally complete arrangement of parts, initiating devices, relays, power supply, and associated components to automatically trip a municipal transmitter or master box over electrical circuits that are electrically isolated from the municipal system circuits.

Parallel Telephone Type. An auxiliary system connected by a municipally controlled individual circuit to the protected property to interconnect the initiating devices at the protected premises and the municipal fire alarm switchboard.

Shunt Auxiliary Type. An auxiliary system electrically connected to an integral part of the municipal alarm system extending the municipal circuit into the protected premises to interconnect the initiating devices, which, when operated, open the municipal circuit shunted around the trip coil of the municipal transmitter or mater box, which is thereupon energized to start transmission without any assistance whatsoever form a local source of power.

Auxiliary Trip Relay. A relay used to operate a municipal master box from an auxiliarized control panel.

Average Ambient Sound Level. The root mean square, A-weighted sound pressure level measured over a 24-hour period.

Bell. A single stroke or vibrating type audible notification appliance which has a bell tone.

Bell, Single Stroke. A bell whose gong is struck only once each time operating energy is applied.

Bell, Vibrating. A bell that rings continuously as long as operating power is applied.

Bell, Motorized. A bell whose gong is motorized and rings continuously as long as operating power is applied.

BOCA. Building Officials and Code Administrators, 4051 W Flossmoor Rd, Country Club Hills, IL 60478.

Box (or Station) Fire Alarm (Noncoded). A manually operated device which, when operated, closes or opens one or more sets of contacts and generally locks the contacts in the operated position until the box is reset.

Box (or Station) Fire Alarm (Coded). A manually operated device in which the act of pulling a lever causes the transmission of not less than three rounds of coded alarm signals. Similar to the noncoded type, except that instead of a manually operated switch, a mechanism to rotate a code wheel is utilized. Rotation of the code wheel, in turn, causes an electrical circuit to be alternately opened and closed, or closed and opened, thus sounding a coded alarm that identifies the location of the box. The code wheel is cut for the individual code to be transmitted by the device and can operate by clockwork or by an electric motor. Clockwork transmitters can be prewound or can be wound by the pulling of the alarm lever. Usually the box is designed to repeat its code four times and automatically come to rest. Prewound transmitters must sound a trouble signal when they require rewinding. Solid state electronic coding devices are also used in conjunction with the fire alarm control unit to produce coded sounding of the audible signaling appliances.

Box, Breakglass Fire Alarm. A fire alarm box in which it is necessary to break a special glass or fiber element in order to operate the box. The special element must be replaced once the box is actuated.

Box, Non-breakglass (BNG) Fire Alarm. A fire alarm box, which is held in its normal state by a spring or other mechanical device. Once the box is actuated, it must be manually reset with a key or special tool.

Box Battery. The battery supplying power for an individual fire alarm box where radio signals are used for the transmission of box alarms.

Bridging Point. The point at which the distribution of signaling line circuits to trunk facilities, leg facilities, or both occurs.

Building, Existing. Any structure erected or officially authorized prior to the effective date of the adoption of this edition of the *Code* by the agency or jurisdiction. With respect to judging whether a building should be considered existing, the deciding factor is not when the building was designed or when construction started but rather the date plans were approved for construction by the appropriate authority having jurisdiction. It is intended that the initial assessment of the building, when new, should be based on new occupancy requirements for the edition of the *Code* in effect on the date of plan approval. Subsequent assessments of the building made while that edition of the *Code* is still in effect should also be based on new occupancy requirements.

Business Occupancy. Business occupancies are those used for the transaction of business (other than those covered under *Mercantile Occupancy*), for the keeping of accounts and records, and for similar purposes. Business occupancies include air traffic control towers (ACTCs); courthouses; city halls; dentists' offices; college and university instructional buildings, classrooms under 50 persons, and instructional laboratories; doctors' offices; general offices; outpatient clinics, ambulatory; and town halls.

Cabinet. Fire alarm control unit enclosure.

CABO. Council of American Building Officials, 5203 Leesburg Pike, Falls Church, VA 22041.

Carrier. High frequency energy that can be modulated by voice or signaling impulses.

Carrier System. A means of conveying a number of channels over a single path by modulating each channel on a different carrier frequency and demodulating at the receiving point to restore the signals to their original form.

Cathode Ray Tube (CRT). An electronic tube used to display data, such as a monitor, that is coated internally with a phosphorescent material. When the electron beam strikes the phosphor, an electron is in turn released thus causing it to glow.

Ceiling. The upper surface of a space, regardless of height. Areas with a suspended ceiling have two ceilings, one visible from the floor and one above the suspended ceiling.

Ceiling Height. The height from the continuous floor of a room to the continuous ceiling of a room or space.

Ceiling Surfaces. Ceiling surfaces referred to in conjunction with the locations of initiating devices are defined as follows:

> **Beam Construction.** Ceilings having solid structural or solid nonstructural members projecting down from the ceiling surface more than 4 in. (100 mm) and spaced more than 3 ft (0.9 m) center to center.

> **Girder.** A support for beams or joists that runs at right angles to the beams or joists. Where the top of girders are within 4 in. (100 mm) of the ceiling, they are a factor in determining the number of detectors and are to be considered as beams. Where the top of the girder is more than 4 in. (100 mm) from the ceiling, it is not a factor in detector locations.

Central Processing Unit (CPU). An arrangement of circuitry using computer circuit techniques usually consisting of memory elements, signal processing circuitry, and a means to input and output data at very high speed.

Central Station. A supervising station that is listed for central station service.

Central Station Fire Alarm System. A system or group of systems in which the operations of circuits and devices are transmitted automatically to, recorded in, maintained by, and supervised from a listed central station having competent and experienced servers and operators who, upon receipt of a signal, take such action as required by this code. Such service is to be controlled and operated by a person, firm, or corporation whose business is the furnishing, maintaining, or monitoring of supervised fire alarm systems.

Central Station Service. The use of a system or a group of systems in which the operations of circuits and devices at a protected property are signaled to, recorded in, and supervised from a listed central station having competent and experienced operators who, upon receipt of a signal, take such action as required by NFPA 72, *National Fire Alarm Code*. Related activities at the protected property such as equipment installation, inspection, testing, maintenance, and runner service are the responsibility of the central station or a listed fire alarm service-local company. Central station service is controlled and operated by a person, firm, or corporation whose business is the furnishing of such contracted services or whose properties are the protected premises.

Certification. A systematic program using randomly selected follow-up inspections of the certified systems installed under the program that allows the listing organization to verify that a fire alarm system complies with all the requirements of this code. A system installed under such a program is identified by the issuance of a certificate and is designated as a certificated system.

Certification of Personnel. A formal program of related instruction and testing as provided by a recognized organization or the authority having jurisdiction.

Channel. A path for voice or signal transmission utilizing modulation of light or alternating current within a frequency band.

Chime. A single-stroke or vibrating-type audible signal appliance that has a xylophone-type striking bar.

Circuit. The conductors or radio channel and associated equipment used to perform a definite function in connection with an alarm system.

Circuit Interface. A circuit component that interfaces initiating devices or control circuits, or both, notification appliances or circuits, or both, system control outputs, and other signaling line circuits to a signaling line circuit.

Circuit Resistance. The total resistance measured at the termination point of a circuit, excluding any connected components.

Class A Circuit. Circuits capable of transmitting an alarm signal during a single open or a nonsimultaneous single ground fault on a circuit conductor shall be designated as Class A. It is commonly known as a 4-wire circuit.

Class B Circuit. Circuits not capable of transmitting an alarm beyond the location of a single open or a nonsimultaneous single ground fault on a circuit conductor shall be designated as Class B. It is commonly known as a 2-wire circuit with end-of-line resistor.

Cloud Chamber Smoke Detection. The principle of using an air sample drawn from the protected area into a high humidity chamber combined with a lowering of chamber pressure to create an environment in which the resultant moisture in the air condenses on any smoke particles present, forming a cloud, The cloud density is measured by a photoelectric principle. The density signal is processed and used to convey an alarm condition when it meets preset criteria.

Coded. An audible or visible signal conveying several discrete bits or units of information. Notification signal examples are numbered strokes of an impact-type appliance and numbered flashes of a visible appliance.

Coded Signal. A signal pulsed in a prescribed code for each round of transmission. A minimum of three rounds and a minimum of three impulses are required for an alarm signal.

Combination Detector. A device that either responds to more than one of the fire phenomenon or employs more than one operating principle to sense one of these phenomenon. Typical examples are a combination of a heat detector with a smoke detector or a combination rate-of-rise and fixed-temperature heat detector.

Combination Fire Alarm and Guard's Tour Box. A manually operated box for separately transmitting a fire alarm signal and a distinctive guard patrol tour supervisory signal.

Combination System. A fire alarm system whose components might be used, in whole or in part, in common with a nonfire signaling system, such as a paging system, a security system, a building automatic system, or a process monitoring system.

Combustible. Capable of undergoing combustion.

Combustion. A chemical process that involves oxidation sufficient to produce light or heat.

Communications Channel. A circuit or path connecting a subsidiary station(s) over which signals are carried.

Compatibility Listed. A specific listing process that applies only to two-wire devices (such as smoke detectors) designed to operate with certain control equipment.

Compatible (Equipment). Equipment that interfaces mechanically or electrically as manufactured without field modification.

Contiguous Property. A single-owner or single-user protected premises on a continuous plot of ground, including any buildings thereon, that is not separated by a public thoroughfare, transportation right-of-way, property owned or used by others, or body of water not under the same ownership.

Control-by-Event. A method of providing a variety of output responses based upon various initiating conditions (events).

Control Unit. A system component that monitors inputs and controls outputs through various types of circuits.

Conventional Control Panel. A fire alarm panel that can only monitor on/off signals.

Cross Zone. An initiating circuit configuration, which allows an alarm only after two or more devices, is in alarm.

Critical Radiant Flux. The level of incident radiant heat energy on a floor covering system at the most distant flameout point as determined by the test procedure of NFPA 253, *Standard Method of Test for Critical Radiant Flux of Floor Covering Systems Using a Radiant Heat Energy Source.* The unit of measurement of critical radiant flux is watts per square centimeter (W/sq cm).

Data Gathering Panel (DGP). A remote fire alarm control unit that sends information to a master control unit.

Day-Care Occupancy. Day-care occupancies include all buildings or portions of buildings in which four or more clients receive care, maintenance, and supervision, by other than their relatives or legal guardians, for less than 24 hours per day. Day-care occupancies include; child day-care occupancies; kindergarten classes that are incidental to a child day-care occupancy; adult day-care occupancies, except where part of a health care occupancy; nursery schools; and day-care homes.

Day/Night Sensitivity. A method by which intelligent detectors are forced into high or low sensitivity at certain times of the day.

Delinquency Signal. A signal indicating the need for action in connection with the supervision of guards or system attendants.

Deluge System. A system which releases water in the event of a fire.

Derived Channel. A signaling line circuit that uses the local leg of the public switched network as an active multiplex channel while simultaneously allowing that leg's use for normal telephone communications.

Detector. A device suitable for connection to a circuit having a sensor that responds to a physical stimulus such as heat or smoke.

Detector Coverage. The recommended area that a detector is listed to protect.

Detention and Correction Occupancy. Detention and correctional occupancies are used to house individuals under varied degrees of restraint or security and are occupied by persons who are mostly incapable of self-preservation because of security measures not under the occupants' control. Detention and correctional occupancies include; adult and juvenile substance abuse centers; adult local detection facilities; adult and juvenile work camps; juvenile community residential centers; adult community   residential centers; juvenile detention facilities; adult correctional institutions; and juvenile training schools.

Digital Alarm Communicator Receiver (DACR). A system component that accepts and display signals from digital alarm communicator transmitters (DACTs) sent over the public switched telephone network.

Digital Alarm Communicator System (DACS). A system in which signals are transmitted from a digital alarm communicator transmitter (DACT) located at the protected premises through the public switched telephone network to a digital alarm communicator receiver (DACR).

Digital Alarm Communicator Transmitter (DACT). A system component at the protected premises to which initiating devices or groups of devices are connected. The DACT seizes the connected telephone line, dials a preselected number to connect to a DACR, and transmits signals indicating a status change of the initiating device.

Digital Alarm Radio Receiver (DARR). A system component composed of two subcomponents: one that receives and decodes radio signals, and another that annunciates the decoded data. These two subcomponents can be coresident at the central station or separated by means of a data transmission channel.

Digital Alarm Radio System (DARS). A system in which signals are transmitted from a digital alarm radio transmitter (DART) located at a protected premises through a radio channel to a digital alarm radio receiver (DARR).

Digital Alarm Radio Transmitter (DART). A system component that is connected to or an integral part of a DACT that is used to provide an alternate radio transmission channel.

Diode Matrix. A circuit of diodes that allows the fire alarm control unit to control outputs by inputs using logic, such as Floor Above, Floor Below, and Fire Floor. A diode matrix is typically found in hard-wired panels, which require such features. In microprocessor-based panels, software programming replaces the diode matrix.

Display. The visual representation of output data as opposed to printed copy.

Door Holder. An electromagnetic device used to hold a door open while energized. In the event of a fire, the door holder is de-energized and the door is allowed to close.

Double Doorway. A single opening that has no intervening wall space or door trim separating the two doors.

Draft Stop. A continuous membrane used to subdivide a concealed space to restrict the passage of smoke, heat, and flames.

Drift Compensation. An algorithm which permits a smoke detector to maintain a constant sensitivity setting by adjusting for environmental contaminants and other factors.

Dual Control. The use of two primary trunk facilities over separate routes or different methods to control one communications channel.

Duct Detector. Initiating devices designed to sample air passing through an HVAC duct and allows for early detection of a developing fire.

Dwelling Unit. A single unit, providing complete, independent living facilities for one or more persons, including permanent provisions for living, sleeping, eating, cooking, and sanitation.

Educational Occupancy. Educational occupancies include all buildings or portions of buildings used for educational purposes through the twelfth grade by six or more persons for four or more hours per day or more than 12 hours per week. Educational occupancies include academies, nursery schools, kindergartens, and schools.

EIA-232 and EIA-485. Communications interface standards as described by the Electronics Industries Association, a US standards making organization.

Ember. A particle of solid material that emits radiant energy due either to its temperature or the process of combustion on its surface. (See *Spark*). Class A and Class D combustibles burn as embers under conditions where the flame typically associated with fire does not necessarily exist. This glowing combustion yields radiant emissions in parts of the radiant energy spectrum that are radically different from those parts affected by flaming combustion. Specialized detectors, specifically designed to detect those emissions, should be used in applications where this type of combustion is expected. In general, flame detectors are not intended for the detection of embers.

Emergency Voice/Alarm Communications. Dedicated manual or automatic facilities for originating and distributing voice instructions, as well as alert and evacuation signals pertaining to a fire emergency, to the occupants of a building.

End-of-Line Device. A device used to terminate a supervised circuit. Typically a resistor or diode is installed in series at the end of a two-wire circuit to maintain supervision.

End-of-Line Relay. A relay used to supervise power, for example, supplied to a 4-wire smoke detector or power to an addressable device. The relay is installed within or adjacent to the last powered device on the circuit.

Erasable Programmable Read Only Memory (EPROM). A nonvolatile semiconductor memory component whose contents may be erased, usually through exposure to ultraviolet light. EPROMs are used to store digital data.

Evacuation. The withdrawal of occupants from a building. Evacuation does not include relocation of occupants within a building.

Evacuation Signal. A distinctive signal intended to be recognized by the occupants as requiring evacuation of the building.

Existing. That which is already in existence on the date when the current edition of the *Code* goes into effect. See *Building, Existing*.

Exit. That portion of a means of egress that is separated from all other spaces of the building or structure by construction or equipment as required in NFPA 101 5-1.3.2.1 to provide a protected way of travel to the exit discharge. Exits include exterior exit doors, exit passageways, horizontal exits, separated exit stairs, and separated exit ramps.

Exit Access. That portion of a means of egress that leads to an exit.

Exit Plan. A plan for the emergency evacuation of the premises.

External Interface. An EIA-485 bi-directional serial port used to upload and download data.

Family Living Unit. One or more rooms in a single-family detached dwelling, single-family attached dwelling, multifamily dwelling, or mobile home for the use of one or more persons as a housekeeping unit with space for eating, living, and sleeping and permanent provisions for cooking and sanitation. This definition covers living areas only and not common usage areas in multifamily dwellings such as corridors, lobbies, or basements.

Fault. An open, ground, or short condition on any lines(s) extending from a control unit, which could prevent normal operation.

Field of View. The solid cone extending out from the detector within which the effective sensitivity of the detector is at least 50 percent of its on-axis, listed, or approved sensitivity.

Fire. A chemical reaction that occurs when a combustible material is exposed to oxygen during which rapid oxidation results in the release of heat, light, flame, and/or smoke.

Fire Alarm Control Unit (Panel). A system component that receives inputs from automatic and manual fire alarm devices and might supply power to detection devices and a transponder(s) or an off-premises transmitter(s). The control unit might also provide transfer of power to the notification appliances and transfer of condition to relays or devices connected to the control unit. The fire alarm control unit can be a local fire alarm control unit or master control unit.

Fire Alarm/Evacuation Signal Tone Generator. A device that, upon command, produces a fire alarm/evacuation tone.

Fire Alarm Signal. A signal initiated by a fire alarm-initiating device such as a manual fire alarm box, automatic fire detector, waterflow switch, or other device whose activation is indicative of the presence of a fire or fire signature.

Fire Alarm System. A system or portion of a combination system consisting of components and circuits arranged to monitor and annunciate the status of fire alarm or supervisory signal-initiating devices and to initiate the appropriate response to those signals.

Fire Barrier. A continuous membrane, either vertical or horizontal, such as a wall or floor assembly that is designed and constructed with a specified fire resistance rating to limit the spread of fire and that also will restrict the movement of smoke. Such barriers might have protected openings.

Fire Barrier Wall. A wall, other than a fire wall, having a fire resistance rating.

Fire Command Center. The principal attended or unattended location where the status of the detection, alarm communications, and control systems is displayed and from which the system(s) can be manually controlled.

Fire Compartment. A space within a building, that is enclosed by fire barriers on all sides, including the tip and bottom. In the provisions of fire compartments utilizing the outside walls of a building, it is not intended that the outside wall be specifically fire resistance-rated unless required by other standards. Likewise it is not intended for outside windows or doors to be protected unless specifically required for exposure protection by another section of this *Code* or by other standards.

Fire Protection Rating. The designation indicating the duration of the fire test exposure to which a fire door assembly or fire window assembly was exposed and successfully met all the acceptance criteria as determined in accordance with NFPA 252, *Standard Methods of Fire Tests of Door Assemblies*, or NFPA 257, *Standard on Fire Test for Window and Glass Block Assemblies*, respectively.

Fire Rating. The classifications indicating in time (hours) the ability of a structure or component to withstand fire conditions.

Fire Resistance Rating. The time, in minutes or hours, that materials or assemblies have withstood a fire exposure as established in accordance with the test procedures of NFPA 251, *Standard Methods of Tests of Fire Endurance of Building Construction and Materials*.

Fire-Gas Detector. A device that detects the gases produced by a fire.

Fire Safety Function Control Device. The fire alarm system component that directly interfaces with the control system that controls the fire safety function.

Fire Safety Functions. Building and fire control functions that are intended to increase the level of life safety for occupants or to control the spread of the harmful effects of fire.

Fire Warden. A building staff member or a tenant trained to perform assigned duties in the event of a fire emergency.

Fixed Temperature Detector. A device that responds when its operating element becomes heated to a predetermined level. The difference between the operating temperature of a fixed temperature device and the surrounding air temperature is proportional to the rate at which the temperature is rising and is commonly referred to as "thermal lag." The air temperature is always higher than the operating temperature of the device. Typical examples of fixed temperature-sensing elements follow.

> *(a) Bimetallic.* A sensing element comprised of two metals having different coefficients of thermal expansion arranged so that the effect is deflection in one direction when heated and in the opposite direction when cooled.
>
> *(b) Electrical Conductivity.* A line-type or spot-type sensing element whose resistance varies as a function of temperature.
>
> *(c) Fusible Alloy.* A sensing element of a special composition (eutectic) metal that melts rapidly at the rated temperature.
>
> *(d) Heat-Sensitive Cable.* A line-type device whose sensing element comprises, in one type, two current-carrying wires separated by heat-sensitive insulation that softens at the rated temperature, thus allowing the wires to make electrical contact. In another type, a single wire is centered in a metallic tube, and the intervening space is filled with a substance that, at a critical temperature, becomes conductive, thus establishing electrical contact between the tube and the wire.
>
> *(e) Liquid Expansion.* A sensing element comprising a liquid capable of marked expansion in volume in response to temperature increase.

Flame. A body or stream of gaseous material involved in the combustion process and emitting radiant energy at specific wavelength bands determined by the combustion chemistry of the fuel. In most cases, some portion of the emitted radiant energy is visible to the human eye.

Flame Detector. A radiant energy-sensing fire detector that detects the radiant energy emitted by a flame. Flame detectors are categorized as ultraviolet, single wavelength infrared, ultraviolet infrared, or multiple wavelength infrared.

Flame Detector Sensitivity. The distance along the optical axis of the detector at which the detector can detect a fire of specified size and fuel within a given time frame.

Flame Spread. The propagation of flame over a surface.

FM. Factory Mutual Engineering Corporation, 1151 Boston Providence Turnpike, Norwood, MA 02062.

Four-Wire Smoke Detector. A smoke detector which is connected to the fire alarm control unit by two two-wire circuits. One circuit is the initiating device circuit while the other circuit provides supervised power to the detector, typically 12 VDC or 24 VDC.

Frequency Division Multiplexing. A signaling method characterized by the simultaneous transmission of more than one signal in a communication channel. Signals from one or multiple terminal locations are distinguished from one another by virtue of each signal being assigned to a separate frequency or combination of frequencies.

General Alarm. A term usually applied to the simultaneous operation of all audible alarm signals on a system to indicate the need to evacuate a building.

Ground Fault. A condition in which the resistance between a conductor and ground reaches an unacceptably low level.

Ground Fault Detector. Detects the presence of a ground condition on system wiring.

Guard Signal. A supervisory signal monitoring the performance of guard patrols.

Guard's Tour Reporting Station. A device that is manually or automatically initiated to indicate the route being followed and the timing of a guard's tour.

Guard's Tour Supervision. Devices that are manually or automatically initiated to indicate the route being followed and the timing of a guard's tour.

Hazardous Area. Those areas of structures or buildings posing a degree of hazard greater than that normal to the general occupancy of a building or structure, such as those areas used for the storage or use of combustibles or flammables; toxic, noxious, or corrosive materials; or heat-producing appliance.

Health Care Occupancy. Health care occupancies are those used for purposes such as medical or other treatment or care of persons suffering from physical or mental illness, disease, or infirmity; and for the care of infants, convalescents, or infirm aged persons. Health care occupancies provide sleeping facilities for four or more occupants and are occupied by persons who are mostly incapable of self-preservation because of age, physical or mental disability, or because of security measures not under the occupants' control. Health care occupancies include hospitals, limited care facilities, and nursing homes. Health care occupancies also include ambulatory health care centers.

Heat Alarm. A single or multiple station alarm responsive to heat.

Heat Detector. A fire detector that senses heat produced by burning substances. Heat is the energy produced by combustion that causes substances to rise in temperature.

High-Rise Building. A building more than 75 ft (23 m) in height. Building height shall be measured from the lowest level of fire department vehicle access to the floor of the highest occupiable story. It is the intent of this definition that in determining the level from which the highest occupiable floor is to be measured, the enforcing agency should exercise reasonable judgment, including consideration of overall accessibility to the building by fire department personnel and vehicular equipment. Where a building is situated on a sloping terrain and there is building access on more than one level, the enforcing agency may select the level that provides the most logical and adequate fire department access.

Horn. An audible notification appliance in which electrical energy is used to produce a sound by driving a device which imparts motion to a flexible component that vibrates at a nominal frequency.

Horn/Strobe. A combination audible and visible notification appliance, which operates as a horn and a strobe light, simultaneously. The horn produces a sound at a nominal frequency and the strobe light flashes at a predetermined rate.

Household. The family living unit in single-family detached dwellings, single-family attached dwellings, multi-family buildings, and mobile homes.

Household Fire Alarm System. A system of devices that produces an alarm signal in the household for the purpose of notifying the occupants of the presence of a fire so that they will evacuate the premises.

Hunt Group. A group of associated telephone lines within which an incoming call is automatically routed to an idle (not busy) telephone line for completion.

ICBO. International Conference of Building Officials, 5360 Workman Mill Road, Whittier, CA 90601.

Industrial Occupancy. Industrial occupancies include factories making products of all kinds and properties devoted to operations such as processing, assembling, mixing, packaging, finishing or decorating, and repairing. Industrial occupancies include dry cleaning plants; laundries; factories of all kinds; power plants; food processing plants; pumping stations; gas plants; refineries; hangars (for servicing/maintenance); sawmills; and telephone exchanges.

Initiating Device. A system component that originates transmission of a change of state condition, such as in a smoke detector, manual fire alarm box, or supervisory switch.

Initiating Device Circuit (IDC). A circuit to which automatic or manual initiating devices are connected where the signal received does not identify the individual device operated.

Intelligent System. A system using analog devices communicating with a control unit that individually monitors the value or status reported by the analog sensors and makes the normal, alarm, or trouble decisions.

Integrated System. A computer-based control system listed for use as a fire alarm system, in which certain components are common to nonfire monitoring and control functions.

Intermediate Fire Alarm or Fire Supervisory Control Unit. A control unit used to provide area fire alarm or area fire supervisory service that, where connected to the proprietary fire alarm system, becomes a part of that system.

Ionization Smoke Detection. The principle of using a small amount of radioactive material to ionize the air between two differentially charged electrodes to sense the presence of smoke particles. Smoke particles entering the ionization volume decrease the conductance of the air by reducing ion mobility. The reduced conductance signal is processed and used to convey an alarm condition when it meets preset criteria. Ionization smoke detection is more responsive to invisible particles (smaller than 1 micron in size) produced by most flaming fires. It is somewhat less responsive to the larger particles typical of most smoldering fire. Smoke detectors utilizing the ionization principle are usually of the spot type.

Labeled. Equipment or materials to which has been attached a label, symbol, or other identifying mark of an organization that is acceptable to the authority having jurisdiction and concerned with product evaluation that maintains periodic inspection of production of labeled equipment or materials and by whose labeling the manufacturer indicates compliance with appropriate standards or performance in a specified manner.

Lamp Test. Activation of all control unit lights for test purposes.

Last Event Recall. The ability to review the last event in a control unit history.

Leg Facility. The portion of a communications channel that connects not more than one protected premises to a primary or secondary trunk facility. The leg facility includes the portion of the signal transmission circuit from its point of connection with a trunk facility to the point where it is terminated within the protected premises at one or more transponders.

Level Ceilings. Ceilings that are actually level or have a slope of 1 _ in./ft (41.7 mm/m) or less.

Lexan®. Single strength glass (which is easily replaced when broken) manufactured by GE plastics used for windows and covers on manual pull stations and on fire alarm cabinet doors.

Light Emitting Diode (LED). A diode that emits visible light when current is applied. LED's are used as single element visible indicators or as multiple-segment displays.

Light Scattering. The action of light reflected and/or refracted off particles of combustion for detection in a photoelectric smoke detector, also known as the "Tyndall Effect".

Limited-Combustible. As applied to a building construction material, other than interior finish, means a material not complying with the definition of noncombustible material that, in the form in which it is used has a potential heat value not exceeding 3500 Btu/lb (8.14×10^6 J/kg) and complies with one of the following paragraphs, (a) or (b).

    (a) Materials having a structural base of noncombustible material with a surface not exceeding a thickness of 1/8 in. (0.3 cm) that has a flame spread rating not greater than 50.

    (b) Materials, in the form and thickness used, other than as described in (a), having neither a flame spread rating greater than 25 nor evidence of continued progressive combustion, and of such composition that surfaces that would be exposed by cutting through the material on any plane would have neither a flame spread rating greater than 25 nor evidence of continued progressive combustion.

Materials subject to increase combustibility or flame spread rating beyond the limits herein established through the effects of age, moisture, or other atmospheric condition shall be considered combustible. See NFPA 259, *Standard Test Method for Potential Heat of Building Materials*, and NFPA 220, *Standard on Types of Building Construction*.

Line-Type Detector. A device in which detection is continuous along a path. Typical examples are rate-of-rise pneumatic tubing detectors, projected beam smoke detectors, and heat-sensitive cable.

Liquid Crystal Display (LCD). An arrangement of individual segments used to display information. Each segment becomes transparent or opaque as electric current is applied or removed from each segment.

Listed. Equipment, materials, or services included in a list published by an organization acceptable to the authority having jurisdiction and concerned with evaluation of products or services that maintains periodic inspection of production of listed equipment or materials or periodic evaluation of services and whose listing states either that the equipment, material, or service meets identified standards or has been tested and found suitable for a specified purpose. The means for identifying listed equipment may vary for each organization concerned with product evaluation, some of which do not recognize equipment as listed unless it is also labels. The authority having jurisdiction should utilize the system employed by the listing organization to identify a listed product.

Living Area. Any normally occupiable space in a residential occupancy, other than sleeping rooms or rooms that are intended for combination sleeping/living, bathrooms, toilet compartments, kitchens, closets, halls, storage or utility spaces, and similar areas.

Loading Capacity. The maximum number of discrete elements of fire alarm systems permitted to be used in a particular configuration.

Local Control Unit (Panel). A control unit that serves the protected premises or a portion of the protected premises and indicates the alarm via notification appliances inside the protected premises.

Local Energy Master Box. A municipal master box that uses electrical energy from the protected premises to energize its electromagnetic tripping mechanism. See *Shunt Master Box*.

Local Fire Alarm System. A local system sounding an alarm at the protected premises as a result of the manual operation of a fire alarm box or the operation of protection equipment or systems, such as water flowing in a sprinkler system, the discharge of carbon dioxide, the detection of smoke, or the detection of heat.

Local Supervisory System. A local system arranged to supervise the performance of guard patrols or the operative condition of an automatic sprinkler system or other system used for the protection of life and property against fire.

Loss of Power. The reduction of available voltage at the load below the point at which equipment can function as designed.

Low Power Radio Transmitter. Any device that communicates with associated control/receiving equipment by low power radio signals.

Maintenance. Repair service, including periodic inspections and tests, required to keep the fire alarm system and its component parts in an operative condition at all times, together with replacement of the system or its components when they become undependable or inoperable for any reason.

Manual Fire Alarm Box. A manually operated device used to initiate an alarm signal.

Master Box. A municipal fire alarm box that can also be operated by remote means.

Master Clock. Time indicating appliance, which is designed to provide, synchronized time reporting for secondary clocks.

Master Control Unit (Panel). A control unit that serves the protected premises or portion of the protected premises as a local control unit and accepts inputs from other fire alarm control units.

Mercantile Occupancy. Mercantile occupancies include stores, markets, and other rooms, buildings, or structures for the display and sale of merchandise. Mercantile occupancies include auction rooms, shopping centers, department stores, supermarkets, and drugstores.

Microprocessor. A complex circuit element that is usually the main control for a central processing unit. A single such component may house the entire central processor unit in a fire alarm control unit.

Model Codes. BOCA National Code, ICBO Uniform Building Code, SBCCI Standard Building Code.

MPS (Main Power Supply). The main power supply of a control unit provides all of the necessary power to operate the control unit plus the power required to operate automatic initiating devices and notification appliances and annunciators.

Multiple Station Alarm. A single station alarm capable of being interconnected to one or more additional alarms so that the actuation of one causes the appropriate alarm signal to operate in all interconnected alarms.

Multiple Station Alarm Device. Two or more single station alarm devices that can be interconnected so that actuation of one causes all integral or separate audible alarms to operate. It also can consist of one single station alarm device having connections to other detectors or to a manual fire alarm box.

Multiplexed Initiating Device Loop. A circuit that connects the transponder or digital alarm communicator transmitter (DACT) to a number of initiating device interfaces.

Multiplexing. A signaling method characterized by simultaneous or sequential transmission, or both, and reception of multiple signals on a signaling line circuit, a transmission channel, or a communications channel, including means for positively identifying each signal.

Municipal Communications Center. The building or portion of a building used to house the central operating part of the fire alarm system; usually the place where the necessary testing, switching, receiving, retransmitting, and power supply devices are located.

Municipal Fire Alarm Box (Street Box). An enclosure housing a manually operated transmitter used to send an alarm to the public fire service communications center.

Municipal Fire Alarm System. A system of alarm-initiating devices, receiving equipment, and connecting circuits (other than a public telephone network) used to transmit alarms from street locations to the public fire service communications center.

Municipal Transmitter. A transmitter that can only be tripped remotely that is used to send an alarm to the public fire service communications center.

Network System. A system where individual control units form a single network and are typically monitored via a CRT.

NFPA. National Fire Protection Association, 1 Batterymarch Park, Quincy, MA 02269.

Noise. An electronics term that covers all types of unwanted electrical signals. Noise signals originate from such sources as fluorescent lamps; walkie-talkies; amateur and CB radios; electronic machines being switched on and off; power surges; and others. Electronic equipment must be able to tolerate increasing amounts of electrical interference. The quality of electronic equipment is directly proportional with its ability to reject and otherwise ignore noise.

Noncoded. An audible or visible signal conveying one discrete bit of information.

Noncombustible. A material that, in the form in which it is used and under the conditions anticipated will not aid combustion or add appreciable heat to an ambient fire. Materials, where tested in accordance with ASTM E136, *Standard Test Method for Behavior of Materials in a Vertical Tube Furnace at 750°C*, and conforming to the criteria contained in Section 7 of the referenced standard shall be considered as noncombustible.

Noncontiguous Property. An owner- or user-protected premises where two or more protected premises, controlled by the same owner or user, are separated by a public thoroughfare, body of water, transportation right-of-way, or property owned by others.

Nonrestorable Initiating Device. A device which contains a sensing element that is designed to be destroyed in the process of operation.

Notification Appliance. A fire alarm system component such as a bell, horn, speaker, light, or text display that provides audible, tactile, or visible outputs or any combination thereof.

> **Audible Notification Appliance.** A notification appliance that alerts by the sense of hearing.
>
> **Audible Textual Notification Appliance.** A notification appliance that conveys a stream of audible information. An example of an audible textual appliance is a speaker that reproduces a voice message.
>
> **Olfactory Notification Appliance.** A notification appliance that alerts by the sense of smell.
>
> **Tactile Notification Appliance.** A notification appliance that alerts by the sense of touch or vibration.
>
> **Visible Notification Appliance.** A notification appliance that alerts by the sense of sight.
>
> **Visible Textual Notification Appliance.** A notification appliance that conveys a stream of visible information. An example of a visible textural appliance is a monitor that displays an alphanumeric or pictorial message.

Notification Appliance Circuit (NAC). A circuit or path directly connected to a notification appliance(s).

Notification Zone. An area covered by notification appliances that are activated simultaneously.

Nuisance Alarm. Any alarm caused by mechanical failure, malfunction, improper installation, or lack of proper maintenance, or any alarm activated by a cause that cannot be determined.

Obscuration. A reduction in the atmospheric transparency caused by smoke, usually expressed as percent per foot.

Occupancy. The purpose for which a building or portion thereof is used or intended to be used.

Occupant Load. The total number of persons that might occupy a building or portion thereof at any one time.

Occupiable Story. A story occupied by people on a regular basis. Stories used exclusively for mechanical equipment rooms, elevator penthouses, and similar spaces are not occupiable stories.

Off-Hook. To make connection with the public switched telephone network in preparing to dial a telephone number.

On-Hook. To disconnect from the public switched telephone network.

Open Area Detection (Protection). Protection of an area such as a room or space with detectors to provide early warning of fire.

Output Circuits. Speakers, telephone, horn, strobe, bell, signal, etc.

Operating Mode, Private. Protection of an area such as a room or space with detectors to provide early warning of fire.

Operating Mode, Public. Audible or visible signaling only to those persons directly concerned with the implementation and direction of emergency action initiating and procedure in the area protected by the fire alarm system.

Operating System Software. The basic operating system software that is alterable only by the equipment manufacturer or its authorized representative. This software is sometimes referred to as "firmware," "BIOS," or "executive program."

Ownership. Any property or building or its contents under legal control by the occupant, by contract, or by holding of a title or deed.

Oxidation. The act or process of oxidizing.

Oxidize. To combine with oxygen. To change a compound by increasing the proportion of the electronegative part or change an element or ion from a lower to a higher positive valence. To remove one or more electrons from an atom, ion or molecule.

Paging System. A system intended to page one or more persons by such means as voice over loudspeaker, coded audible signals, or lamp annunciators.

Parallel Telephone System. A telephone system in which an individually wired circuit is used for each fire alarm box.

Parallel Telephone Auxiliary Alarm System. See *Auxiliary Fire Alarm System.*

Path (Pathways). Any conductor, optic fiber, radio carrier, or other means for transmitting fire alarm system information between two or more locations.

Particles of Combustion. Substances resulting from the chemical process of fire. Such particles include ash which remains at the site of the fire and volatile products, which may evaporate, agglomerate, or condense.

Permanent Visual Record (Recording). An immediately readable, not easily alterable, print, slash, or punch record of all occurrence of status changes.

Photoelectric Light Obscuration Smoke Detection. The principle of utilizing a light source and a photosensitive sensor onto which the principal portion of the source emissions is focused. When smoke particles enter the light path, some of the light is scattered and some is absorbed, thereby reducing the light reaching the receiving sensor. The light reduction signal is processed and used to convey an alarm condition when it meets preset criteria. The response of photoelectric light obscuration smoke detectors is usually not affected by the color of smoke. Smoke detectors utilizing the light obscuration principle are usually of the line type. These detectors are commonly refered to as "projected beam smoke detectors."

Photoelectric Light-Scattering Smoke Detection. The principle of utilizing a light source and a photosensitive sensor arranged in a manner so that the rays from the light source do not normally fall onto the photosensitive sensor. When smoke particles enter the light path, some of the light is scattered by reflection and refraction onto the sensor. The light signal is processed and used to convey an alarm condition when it meets preset criteria. Photoelectric light-scattering smoke detection is more responsive to visible particles (larger than 1 micron in size) produced by most smoldering fires. It is somewhat less responsive to the smaller particles typical of most flaming fires.

Plant. One or more buildings under the same ownership or control on a single property.

Plenum. An air compartment or chamber to which one or more ducts are connected and that forms part of an air distribution system.

Positive Alarm Sequence. An automatic sequence that results in an alarm signal even when manually delayed for investigation, unless the system is reset.

Positive Non-Interfering (PNI) System. A system in which the alarm initiating devices are electrically arranged such that if one device is operated to transmit an alarm, no other devices connected to the same circuit are allowed to interfere with the transmission of a complete and unmodified round of alarm signals.

Positive Non-Interfering and Successive System. A system which prevents subsequently operated initiating devices from interfering with the transmission of an alarm from an operated device and which permits any device to transmit, in turn, its assigned number of rounds of coded signals once the previously actuated device completes its transmission of coded signals.

Power Limited Circuit. A circuit that limits power to the load inherently or by overcurrent protection.

Power Supply. A source of electrical operating power including the circuits and terminations connecting it to the dependent system components.

Presignal System. A feature that allows initial fire alarm signals to sound only in department offices, control rooms, fire brigade stations, or other constantly attended central locations and for which human action is subsequently required to activate a general alarm, or a feature that allows the control equipment to delay the general alarm by more than 1 minute after the start of the alarm processing.

Primary Battery (Dry Cell). A nonrechargeable battery requiring periodic replacement.

Primary Trunk Facility. That part of a transmission channel connecting all leg facilities to a supervising or subsidiary station.

Prime Contractor. The one company contractually responsible for providing central station services to a subscriber as required by this code. This can be either a listed central station or a listed fire alarm service-local company.

Private Radio Signaling. A radio system under control of the proprietary supervising station.

Projected Beam-Type Detector. A type of photoelectric light obscuration smoke detector wherein the beam spans the protected area.

Proprietary Supervising Station. A location to which alarm or supervisory signaling devices on proprietary fire alarm systems are connected and where personnel are in attendance at all times to supervise operation and investigate signals.

Proprietary Supervising Station Fire Alarm System. An installation of fire alarm systems that serves contiguous and noncontiguous properties, under one ownership, from a proprietary supervising station located at the protected property, at which trained, competent personnel are in constant attendance. This includes the proprietary supervising station; power supplies; signal-initiating devices; initiating device circuits; signal notification appliances; equipment for the automatic, permanent visual recording of signals; and equipment for initiating the operation of emergency building control services.

Protected Premises. The physical location protected by a fire alarm system.

Protected Premises (Local) Control Unit (Panel). A control unit that serves the protected premises or a portion of the protected premises and indicated the alarm via notification appliances inside the protected premises.

Protected Premises (Local) Fire Alarm System. A protected premises system that sounds an alarm at the protected premises as the result of the manual operation of a fire alarm box or the operation of protection equipment or system, such as water flowing in a sprinkler system, the discharge of carbon dioxide, the detection of smoke, or the detection of heat.

Public Fire Service Communications Center. The building or portion of the building used to house the central operating part of the fire alarm system; usually the place where the necessary testing, switching, receiving, transmitting, and power supply devices are located.

Public Switched Telephone Network. An assembly of communication facilities and central office equipment operated by authorized common carriers that provides the general public with the ability to establish communication channels via discrete dialing codes.

Radiant Energy Sensing Fire Detector. See *Automatic Fire Detector.*

Radio Alarm Central Station Receiver (RARSR). A system component that receives radio signals. This component resides at a repeater station that is located at a remote receiving location.

Radio Alarm Supervising Station Receiver (RASSR). A system component that receives data and annunciates that data at the supervising station.

Radio Alarm System (RAS). A system in which signals are transmitted from a radio alarm transmitter (RAT) located at a protected premises through a radio channel to two or more radio alarm repeater station receivers (RARRSR) and are annunciated by a radio alarm supervising station receiver (RASSR) located at the central station.

Radio Alarm Transmitter (RAT). A system component at the protected premises to which initiating devices or groups of devices are connected. The RAT transmits signals indicating a status change of the initiating devices.

Radio Channel. A band of frequencies of a width sufficient to allow its use for radio communications. The width of the channel depends on the type of transmissions and the tolerance for the frequency of emission. Channels normally are allocated for radio transmission in a specified type for service by a specified transmitter.

Rate Compensation Detector. A device that responds when the temperature of the air surrounding the device reaches a predetermined level, regardless of the rate of temperature rise. A typical example is a spot-type detector with a tubular casing of a metal that tends to expand lengthwise as it is heated and an associated contact mechanism that closes at a certain point in the elongation. A second metallic element inside the tube exerts an opposing force on the contacts, tending to hold them open. The forces are balanced in such a way that, on a slow rate-of-temperature rise, there is more time for heat to penetrate to the inner element, which inhibits contact closure until the total device has been heated to its rated temperature level. However, on a fast rate-of-temperature rise, there is not as much time for heat to penetrate to the inner element, which exerts less of an inhibiting effect so that contact closure is achieved when the total device has been heated to a lower temperature. This, in effect, compensates for thermal lag.

Rate-of-Rise Detector. A device that responds when the temperature rises at a rate exceeding a predetermined value. Typical examples of rate-of-rise detectors follow.

> *(a) Pneumatic Rate-of Rise Tubing.* A line-type detector comprising small-diameter tubing, usually copper, that is installed on the ceiling or high on the walls throughout the protected area. The tubing is terminated in a detector unit containing diaphragms and associated contacts set to actuate at a predetermined pressure. The system is sealed except for calibrated vents that compensate for normal changes in temperature.
>
> *(b) Spot-Type Pneumatic Rate-of-Rise Detector.* A device consisting of an air chamber, a diaphragm, contacts, and a compensating vent in a single enclosure. The principle of operation is the same as that described for pneumatic rate-of-rise tubing.
>
> *(c) Thermoelectric Effect Detector.* A device whose sensing element comprises a thermocouple or thermopile unit that produces an increase in electric potential in response to an increase in temperature. This potential is monitored by associated control equipment, and an alarm is initiated when the potential increases at an abnormal rate.
>
> *(d) Electrical Conductivity-Type Rate-of-Rise Detector.* A line-type or spot-type sensing element whose resistance changes due to a change in temperature. The rate of change of resistance is monitored by associated control equipment, and an alarm is initiated when the rate of temperature increase exceeds a preset value.

Record Drawings. Drawings (as-built) that document the location of all devices, appliances, wiring sequences, wiring methods, and connections of the components of the fire alarm system as installed.

Record of Completion. A document that acknowledges the features of installation, operation (performance), service, and equipment with representation by the property owner, system installer, system supplier, service organization, and the authority having jurisdiction.

Rectifier. An electrical device without moving parts that changes alternating current to direct current.

Relocation. The movement of occupants from a fire zone to a safe area within the same building.

Remote Supervising Station Fire Alarm System. A system installed in accordance with this code to transmit alarm, supervisory and trouble signals from one or more protected premises to a remote location at which appropriate action is taken.

Repeater Facility. Equipment needed to relay signals between supervisory stations, subsidiary stations, and protected premises.

Repeater Station. The location of the equipment needed for a repeater facility.

Reset. A control function that attempts to return a system or device to its normal, nonalarm state.

Residential Occupancy. Residential occupancies are those occupancies in which sleeping accommodations are provided for normal residential purposes and include all buildings designed to provide sleeping accommodations except for those occupancies classified under health care or detention and correctional occupancies. Residential occupancies include hotels, motels and dormitories; apartment buildings; lodging or rooming houses; one- and two-family dwellings; and board and care facilities.

Restorable Initiating Device. A device whose sensing element is not ordinarily destroyed in the process of operation. Restoration can be manual or automatic.

Retard (Adjustable) Waterflow Switches. An adjustable time delay mechanism in a waterflow switch located between the paddle operated stem and the initiating contacts. This type of switch is designed to prevent brief water flow surges due to water hammer and/or pressure variations from causing an unwanted alarm signal.

Reverse Polarity Interface. A connection from a control unit to a supervising station which operates using polarity reversal to signal an alarm, supervisory or trouble signals.

Ring-Back Circuit. A circuit which functions much like a two-way switch. When a trouble condition occurs, a trouble buzzer and light are energized. When an acknowledge button is depressed, the trouble buzzer and light are reset. When the trouble condition is corrected, the buzzer and light once again are energized indicating an off-normal condition. Once the acknowledge button is released, the circuit returns to normal.

Runner. A person other than the required number of operators on duty at central, supervising or runner stations (or otherwise in contact with these stations) available for prompt dispatching, when necessary, to the protected premises.

Runner Service. The service provided by a runner at the protected premises, including resetting and silencing of all equipment transmitting fire alarm or supervisory signals to an off-premises location.

Satellite Station. A normally unattended remote location. Interconnection of signal-receiving equipment or communication channels from protected premises, which include circuits connecting central supervising station(s) is accomplished here.

Satellite Trunk. A circuit or path connecting a satellite to its central or proprietary supervising station.

SBCCI. Southern Building Code Congress International, Inc., 900 Montclair Rd, Birmingham, AL 35213.

Scanner. Equipment located at the telephone company wire center that monitors each local leg and relays status changes to the alarm center. Processors and associated equipment might also be included.

Secondary Power. See *Standby Battery*.

Secondary Trunk Facility. That part of a transmission channel connecting two or more, but fewer than all, leg facilities to a primary trunk facility.

Separate Sleeping Area. An area of the family living unit in which the bedrooms (or sleeping rooms) are located. Bedrooms (or sleeping rooms) separated by other use areas, such as kitchens or living rooms (but not bathrooms) are considered as separate sleeping areas.

Shall. Indicates a mandatory requirement.

Shapes of Ceilings. The shapes of ceilings are classified as follows:

> **Sloping Ceiling.** A ceiling having a slope of more than 1 1/2"/ft (41.7 mm/m). Sloping ceilings are further classified as follows:
>
> > **Sloping-Peaked Type.** A ceiling in which the ceiling slopes in two directions from the highest point. Curved or domed ceilings can be considered peaked with the slope figures as the slope of the chord from highest to lowest point.
> >
> > **Sloping-Shed Type.** A ceiling in which the high point is at one side with the slope extending toward the opposite side.
>
> **Smooth Ceiling.** A ceiling surface uninterrupted by continuous projections, such as solid joists, beams, or ducts, extending more than 4 in. (100 mm) below the ceiling surface. Open truss constructions are not considered to impede the flow of fire products unless the upper member in continuous contact with the ceiling projects below the ceiling more than 4 in. (100 mm).

Should. Indicates a recommendation or that which is advised but not required.

Shunt Auxiliary Alarm System. See *Auxiliary Fire Alarm System.*

Signal. A status indication communicated by electrical or other means.

Signaling Line Circuit (SLC). A circuit or path between any combination of circuit interfaces, control units or transmitters over which multiple system input signals or output signals, or both, are carried.

Signaling Line Circuit Interface (SLCI). A system component that connects a signaling line circuit to any combination of initiating devices, initiating device circuits, notification appliances, notification appliance circuits, system control outputs, and other signaling line circuits.

Signaling Line Circuit Loop. The physical wire loop along which addressable input and/or output devices are connected.

Signal Silence. A function which causes participating fire alarm activated notification appliances or other outputs to deactivate without otherwise affecting the overall state of the system.

Signal Transmission Sequence. A DACT that obtains dial tone, dials the number(s) of the DACR, obtains verification that the DACR is ready to receive signals, transmits the signals, and receives acknowledgment that the DACR has accepted that signal before disconnecting (going on-hook).

Silence Inhibit. The ability of a fire alarm control unit to prevent the signal silence button from operating until after a prescribed amount of time.

Single Station Alarm. A detector comprising an assembly incorporating a sensor, control components, and an alarm notification appliance in one unit operated from a power source either located in the unit or obtained at the point of installation.

Single Station Alarm Device. An assembly incorporating the detector, the control equipment, and the alarm-sounding device in one unit operated from a power supply either in the unit or obtained at the point of installation.

Site-Specific Software. Software that defines the specific operation and configuration of a particular system Typically, it defines the type and quantity of hardware modules, customized labels, and specific operating features of a system.

Smoke Alarm. A single or multiple station alarm responsive to smoke.

Smoke Barrier. A continuous membrane, either vertical or horizontal, such as a wall, floor, or ceiling assembly, that is designed and constructed to restrict the movement of smoke. A smoke barrier might or might not have a fire resistance rating. Such barriers might have protected openings.

Smoke Compartment. A smoke compartment is a space within a building enclosed by smoke barriers on all sides, including the top and bottom. In the provision of smoke compartments utilizing the outside walls or the roof of a building, it is not intended that outside walls or roofs or any openings therein be capable of resisting the passage of smoke.

Smoke Detector. A device that detects visible or invisible particles of combustion.

Solenoid. A coil of wire so arranged around a core of air, iron, or other material to transform electrical energy to mechanical energy.

Solid Joist Construction. Refers to ceilings having solid structural or solid nonstructural members projecting down from the ceiling surface for a distance of more than 4 in. (100 mm) and spaced at intervals 3 ft (0.9 m) or less, center to center.

Spacing. A horizontally measured dimension related to the allowable coverage of fire detectors.

Spark. A moving ember. The overwhelming majority of applications involving the detection of Class A and Class D combustibles with radiant energy-sensing detectors involve the transport of particulate solid materials through pneumatic conveyor ducts or mechanical conveyors. It is common in the industries that include such hazards to refer to a moving piece of burning material as a "spark" and to systems for the detection of such fires as "spark detection systems".

Spark/Ember Detector. A radiant energy fire detector that is designed to detect sparks or embers, or both. These devices are normally intended to operate in dark environments and in the infrared part of the spectrum.

Spark/Ember Detector Sensitivity. The number of watts (or fraction of a watt) of radiant power from a point source radiator, applied as a unit step signal at the wavelength of maximum detector sensitivity, necessary to produce an alarm signal from the detector within the specified response time.

Spot-Type Detector. A device whose detecting element is concentrated at a particular location. Typical examples are bimetallic detectors, fusible alloy detectors, certain pneumatic rate-of-rise detectors, certain smoke detectors, and thermoelectric detectors.

Standard Audible Emergency Evacuation Signal. A distinctive three-pulse temporal pattern emergency evacuation signal required by NFPA 72 for all new systems installed after July 1, 1996. For a detailed description of this signal see American National Standards (ANSI) S3.41, *Audible Emergency Evacuation Signal.*

Standby Battery. A battery referred to as the secondary power supply, which is kept charged by the fire alarm control unit or by a separate battery charger. When primary (AC) power fails, the battery supplies power for a limited time. The time required for a standby battery to operate the fire alarm system is defined by NFPA 72-1996 1-5.2.6.

Storage Occupancy. Storage occupancies include all buildings or structures utilized primarily for the storage or sheltering of goods, merchandise, products, vehicles, or animals. Storage occupancies include barns; hangars (for storage only); bulk oil storage; parking structures; cold storage; stables; freight terminals; truck and marine terminals; grain elevators; and warehouses.

Story. The portion of a building included between the upper surface of a floor and the upper surface of the floor or roof next above.

Stratification. The phenomenon where the upward movement of smoke and gases ceases due to the loss of buoyancy.

Subscriber. The recipient of contractual supervising station signal service(s). In case of multiple, noncontiguous properties having single ownership, the term refers to each protected premise or its local management.

Subsidiary Station. A subsidiary station is a normally unattended location that is remote from the supervising station and linked by communication channel(s) to the supervising station. Interconnection of signals on one or more transmission channels from protected premises with a communications channel(s) to the supervising station is performed at this location.

Supervising Station. A facility that receives signals and at which personnel are in attendance at all times to respond to these signals.

Supervision. The term supervised refers to monitoring of the circuit, switch, or device in such a manner that a trouble signal is received when a fault that would prevent normal operation of the system occurs.

Supervisory Service. The service required to monitor performance of guard tours and the operative condition of fixed suppression systems or other systems for the protection of life and property.

Supervisory Signal. A signal indicating the need of action in connection with the supervision of guard tours, the fire suppression systems or equipment, or the maintenance features of related systems.

Supervisory Signal-Initiating Device. An initiating device such as a valve supervisory switch, water level indicator, or low-air pressure switch on a dry-pipe sprinkler system whose change of state signals an off-normal condition and its restoration to normal of a fire protection or life safety system, or a need for action in connection with guard tours, fire suppression systems or equipment, or maintenance features of related systems.

Supplementary. As used in NFPA 72, *National Fire Alarm Code*, supplementary refers to equipment or operations not required by NFPA 72, *National Fire Alarm Code*, and designated as such by the authority having jurisdiction.

Switched Telephone Network. An assembly of communications facilities and central office equipment operated jointly by authorized service providers that provides the general public with the ability to establish transmission channels via discrete dialing.

System Unit. The active subassemblies at the central station utilized for signal receiving, processing, display, or recording of status change signals; a failure of one of these subassemblies causes the loss of a number of alarm signals by that unit.

Tamper Supervisory Device. A device used to monitor the position of a water valve, which isolates a waterflow or water pressure switch.

Time Control Module (TCM). Used with a fire alarm control unit to provide releasing service and pre-signal evacuation

Thermal Lag. See *Rate Compensation Detector*.

Transmission Channel. A circuit or path connecting transmitters to supervising station or subsidiary stations on which signals are carried.

Transmitter. A system component that provides an interface between signaling line circuits, initiating device circuits, or control units and the transmission channel.

Transponder. A multiplex alarm transmission system functional assembly located at the protected premises.

Trouble Signal. A signal initiated by the fire alarm system or device indicative of a fault in a monitored circuit or component.

Trunk Facility. That part of a signaling line circuit connecting two or more leg facilities to the central station supervising station or satellite station.

Trunk Primary Facility. That part of a transmission channel connecting all leg facilities to a central or proprietary supervising station or subsidiary station.

Trunk Secondary Facility. That part of a transmission channel connecting two or more, but less than all, leg facilities to a primary trunk facility.

Two-Way Fire Department Communications System. An electrically supervised telephone system providing private voice communication capability between the command center or central control panel and designated remote locations.

UL. Underwriters Laboratories, Inc., 333 Pfingsten Road, Northbrook, IL 60062.

Unwanted Alarm. See *Nuisance Alarm.*

USDOC. U.S. Department of Commerce, National Bureau of Standards, Washington, D.C. 20234.

Visible Signal. A visible signal is the response to the operation of an initiating device by one or more direct or indirect visible notification appliances. For a direct visible signal, the sole means of notification is by direct viewing of the light source. For an indirect visible signal, the sole means of notification is by illumination of the area surrounding the visible signaling appliance.

Voice Paging Protective Signaling System. A manual or automatic supervised paging system used to transmit a voice message which may include information, instructions, or directions following a fire alarm, on a selective or all-call basis.

Walk Test. A feature that allows a single person to test a fire alarm system.

Waterflow Detector. A device which detects the flow of water in a deluge system.

Waterflow Switch. An assembly approved for the service and so constructed and installed that any flow of water from a sprinkler system equal to or greater than from a single automatic sprinkler of the smallest orifice size installed on the system will result in activation of this switch and subsequently indicate an alarm condition.

WATS (Wide Area Telephone Service). Telephone company service allowing reduced costs for certain telephone call arrangement; it can be in-WATS or 800-number service where calls can be placed from anywhere in the continental U.S. to the called party at no cost to the calling party, or out-WATS, a service whereby, for a flat-rate charge, dependant on the total duration of all such calls, a subscriber can make an unlimited number of calls within a prescribed area from a particular telephone terminal without the registration of individual call charges.

Wavelength. The distance between the peaks of a sinusoidal wave. All radiant energy can be described as a wave having a wavelength. Wavelength serves as the unit of measure for distinguishing between different parts of the spectrum. Wavelengths are measured in microns (μm), or Angstroms (Å). The concept of wavelength is extremely important in selecting the proper detector for a particular application. There is a precise interrelation between the wavelength of light being emitted from a flame and the combustion chemistry producing the flame. Specific subatomic, atomic, and molecular events yield radiant energy of specific wavelengths. For example, ultraviolet photons are emitted as the result of the complete loss of electrons or very large changes in electron energy levels. During combustion, molecules are violently torn apart by the chemical reactivity of oxygen, and electrons are released in the process, recombining at drastically lower energy levels, thus giving rise to ultraviolet rations. Visible radiation is generally the result of smaller changes in electron energy levels within the molecules of fuel, flame intermediates, and products of combustion. Infrared radiation comes from the vibration of molecules or parts of molecules when they are in the superheated state associated with combustion. Each chemical compound exhibits a group of wavelengths at which it is resonant. These wavelengths constitute the chemical's infrared spectrum, which is usually unique to that chemical. This interrelationship between wavelength and combustion chemistry affects the relative performance of various types of detectors with respect to various fires.

Wireless Control Panel. A component that transmits/receives and processes wireless signals.

Wireless Initiating Device. Any initiating device that communicates with associated control/receiving equipment by some kind of wireless transmission path.

Wireless Protection System. A system or a part of a system that can transmit and receive signals without the aid of wire. It may consist of any of the following components:

Wireless Control Panel. A component that transmits/receives and processes wireless signals.

Wireless Repeater. A component used to relay signals between wireless receivers or wireless control panels, or both.

Wireless Radio Linker. A device that receives, verifies, and retransmits coded low-power radio frequency alarm and supervisory signals generated by wireless smoke detectors and other wireless initiating devices.

Wireless Repeater. A component used to relay signals between wireless receivers or wireless control panels, or both.

> NOTE: In section 3-13 of the 1996 edition of NFPA 72, the term "wireless" was replaced with "low power radio" to eliminate potential confusion with other transmission media such as optical fiber cables. In making this change, the committee overlooked updating these definitions.

Wireless Smoke Detector. A smoke detector which communicates with associated control / receiving equipment using radio transmission.

Zone. A defined area within the protected premises. A zone can define an area from which a signal can be received, an area to which a signal can be sent, or an area in which a form of control can be executed.

Table 4-4.4.2.1 Corridor Spacing for Visible Appliances

Corridor Length		Minimum Number of 15-cd
ft	m	Visible Appliances Required
0 – 30	0 – 9.14	1
31 – 130	9.45 – 39.6	2
131 – 230	39.93 – 70	3
231 – 330	70.4 – 100.6	4
331 – 430	100.9 – 131.1	5
431 – 530	131.4 – 161.5	6

Table 2-2.1.1.1 Temperature Classification for Heat-Sensing Fire Detectors

Temperature Classification	Temperature Rating Range		Maximum Ceiling Temperature		Color Code
	°F	°C	°F	°C	
Low	100 - 134	39 - 57	20 below	11 below	Uncolored
Ordinary	135 - 174	58 - 79	100	38	Uncolored
Intermediate	175 - 249	80 - 121	150	66	White
High	250 - 324	122 - 162	225	107	Blue
Extra high	325 - 399	163 - 204	300	149	Red
Very extra high	400 - 499	205 - 259	375	191	Green
Ultra high	500 - 575	260 - 302	475	246	Orange

Table 2-2.4.5.1 Heat Detector Spacing Reduction Based on Ceiling Height

Ceiling Height Above		Up to and Including		Multiply Listed Spacing by
ft	m	ft	m	
0	0	10	3.05	1.00
10	3.05	12	3.66	0.91
12	3.66	14	4.27	0.84
14	4.27	16	4.88	0.77
16	4.88	18	5.49	0.71
18	5.49	20	6.10	0.64
20	6.10	22	6.71	0.58
22	6.71	24	7.32	0.52
24	7.32	26	7.93	0.46
26	7.93	28	8.54	0.40
28	8.54	30	9.14	0.34

Appendix A

Underwriters Laboratories

Publications in Numerical Order

NOTE: (DoD) = Department of Defense
(PE) = Professional Engineer

1	Flexible Metal Conduit	1993
3	Flexible Nonmetallic Tubing for Electrical Wiring(DoD)	1984
4	Armored Cable (DoD)	1986
5	Surface Metal Raceways and Fittings (DoD, PE)	1985
6	Rigid Metal Conduit	1993
8	Foam Fire Extinguishers	1990
9	Fire Tests of Window Assemblies	1989
10A	Tin-Clad Fire Doors	1993
10B	Fire Tests of Door Assemblies	1993
13	Power-Limited Circuit Cables	1990
14B	Sliding Hardware for Standard, Horizontally Mounted Tin-Clad Fire Doors	1993
14C	Swinging Hardware for Standard Tin-Clad Fire Doors Mounted Singly & in Pairs	1993
17	Vent or Chimney Connector Dampers for Oil-Fired Appliances	1988
18	Unlined Fire Hose	1980
19	Lined Fire Hose and Hose Assemblies (DoD)	1992
20	General-Use Snap Switches (DoD)	1986
21	LP-Gas Hose	1991
22	Amusement and Gaming Machines	1987
25	Meters for Flammable & Combustible Liquids and LP-Gas	1990
30	Metal Safety Cans	1991
32	Metal Waste Cans	1981
33	Heat Responsive Links for Fire-Protection Service	1993
38	Manually Actuated Signaling Boxes for use with Fire-Protective Signaling Sys(DoD)	1994
44	Rubber-Insulated Wires and Cables	1991
45	Portable Electric Tools (DoD)	1991
47	Semiautomatic Fire Hose Storage Devices	1993
48	Electric Signs	1988
50	Enclosures for Electrical Equipment	1992
51	Power-Operated Pumps for Anhydrous Ammonia and LP-Gas	1993
55A	Materials for Built-Up Roof Coverings	1993

58   Steel Underground Tanks for Flammable & Combustible Liquids [DoD]
62   Flexible Cord and Fixture Wire [DoD]
65   Wired Cabinets
67   Panelboards
69   Electric-Fence Controllers
70   Septic Tanks, Bituminous Coated Metal
72   Tests for Fire Resistance of Record Protection Equipment
73   Motor-Operated Appliances
79   Power-Operated Pumps for Petroleum Product Dispensing Systems
80   Steel Inside Tanks for Oil-Burner Fuel [DoD]
82   Electric Gardening Appliances
83   Thermoplastic-Insulated Wires and Cables [DoD]
87   Power-Operated Dispensing Devices for Petroleum Products [DoD]
92   Fire Extinguisher and Booster Hose
94   Tests for Flammability of Plastic Materials for Parts in Devices & Appliances
96   Lighting Protection Components
96A   Installation Requirements for Lightning Protection Systems
98   Enclosed and Dead-Front Switches [DoD]
103   Chimneys, Factor-Built Residential Type & Building Heating Appliances [DoD]
104   Elevator Door Locking Devices and Contacts
107   Asbestos-Cement Pipe and Couplings
109   Tube Fittings for Flammable & Combustible Fluids, Refrigeration Svs, & Marine Use
112   Portable Wood Ladders
122   Photographic Equipment
123   Oxy-Fuel Gas Torches
125   Valves for Anhydrous Ammonia & LP-Gas (Other Than Safety Relief)
127   Factory-Build Fireplaces
130   Electric Heating Pads
132   Safety Relief Valves for Anhydrous Ammonia & LP-Gas
136   Pressure Cookers
140   Relocking Devices for Safes and Vaults
141   Garment Finishing Appliances
142   Steel Aboveground Tanks for Flammable & Combustible Liquids
144   Pressure Regulating Valves for LP-Gas
147   Hand-Held Torches for Fuel Gases
147A Nonrefillable (Disposable) Type Fuel Gas Cylinder Assemblies
147B Nonrefillable (Disposable) Type Metal Container Assemblies for Butane
150   Antenna Rotators
153   Portable Electric Lamps
154   Carbon-Dioxide Fire Extinguishers

155   Tests for Fire Resistance of Vault and File Room Doors
157   Gaskets and Seals
162   Foam Equipment and Liquid Concentrates
174   Household Electric Storage Tank Water Heaters
180   Liquid-Level Indicating Gauges & Tank-Filling Signals for Petroleum Products
181   Factory-Made Air Ducts and Air Connectors
181A Closure Systems for Use with Rigid Air Ducts & Air Connectors
183   Manufactured Wiring Systems
184   Portable Metal Ladders
187   X-Ray Equipment
193   Alarm Valves for Fire-Protection Service
194   Gasketed Joints for Ductile-Iron Pipe & Fittings for Fire Protection Service
197   Commercial Electric Cooking Appliances
198B Class H Fuses
198C High-Interrupting-Capacity Fuses, Current-Limited Fuses [DoD]
198D Class K Fuses
198E Class R Fuses [DoD]
198F Plug Fuses
198G Fuses for Supplementary Overcurrent Protection
198H Class T Fuses [DoD]
198L D-C Fuses for Industrial Use
198M Mine-Duty Fuses
199   Automatic Sprinklers for Fire-Protection Service
203   Pipe Hanger Equipment for Fire-Protection Service
207   Refrigerant-Containing Components & Accessories, Nonelectric [PE]
209   Cellular Metal Floor Raceways and Fittings [DoD]
213   Rubber Gasketed Fittings for Fire Protection Service
214   Tests for Flame-Propagation of Fabrics & Films
217   Single and Multiple Station Smoke Detectors
218A Battery Contractors for Use in Diesel Engines Driving Centrifugal Fire Pumps
219   Lined Fire Hose for Interior Standpipes
224   Extruded Insulated Tubing
228   Door Closers-Holders, with or without Integral Smoke Detectors
231   Power Outlets
234   Low Voltage Lighting Fixtures for Use in Recreational Vehicles
242   Nonmetallic Containers for Waste Paper
244A Solid-State Controls for Appliances
246   Hydrants for Fire-Protection Service
250   Household Refrigerators and Freezers
252   Compressed Gas Regulators [DoD]
252A Comperssed Gas Regulator Accessories
260   Dry Pipe & Deluge Valves for Fire-Protection Seervice

262	Gate Valves for Fire-Protection Service [DoD]
263	Fire Tests of Building Construction and Materials
268	Smoke Detectors for Fire Protective Signaling Systems
268A	Smoke Detectors For Duct Application
275	Automotive Glass-Tube Fuses
291	Automated Teller Systems
294	Access Control System Units
296	Oil Burners [DoD]
296A	Waste Oil-Burning Air-Heating Appliances
297	Acetylene Generators Portable, Medium-Pressure
298	Portable Electric Hand Lamps [DoD]
299	Dry Chemical Fire Extinguishers [DoD]
300	Fire Testing/ Fire Extinguishing Systems for Protection of Restaurant Cooking Areas
303	Refrigeration and Air-conditioning Condensing and Compressor Units
305	Panic Hardware
307A	Liquid Fuel-Burning Heating Appliances for Manufactured Homes & Rec. Vehicles
307B	Gas Burning Heating Appliances for Mobile Homes & Recreational Vehicles
310	Electrical Quick-Connect Terminals
311	Roof Jacks for Manufactured Homes and Recreational Vehicles
312	Check Valves for Fire-Protection Services
325	Door, Drapery, Gate, Louver, & Window Operators and Systems [DoD]
330	Hose & Hose Assemblies for Dispensing Gasoline [PE]
331	Strainers for Flammable Fluids and Anhydrous Ammonia
340	Tests for Comparative Flammability of Liquids
343	Pumps for Oil-Burning Appliances
346	Waterflow Indicators for Fire Protective Signaling Systems
347	High Voltage Industrial Control Equipment
351	Rosettes
352	Constant-Level Oil Valves
353	Liquid-Tight Flexible Steel Conduit [DoD]
355	Cord Reels
360	Liquid-Tight Flexible Steel Conduit [DoD]
363	Knife Switch
365	Police Station Connected Burglar Alarm Units & Systems
372	Primary Safety Controls for Gas & Oil-Fired Appliances
378	Draft Equipment
385	Play Pipes for Water Supply Testing in Fire-Protection Services
391	Solid-Fuel & Combination-Fuel Central & Supplementary Furnaces
393	Indicating Pressure Gauges for Fire Protection Services
395	Automotive Fuel Tanks
399	Drinking-Water Coolers [PE]
401	Portable Spray Hose Nozzles for Fire Protection Service

404	Gauges, Indicating Pressure, for Compressed Gas Service
405	Fire Department Connections
407	Manifolds for Compressed Gases
408	Acetylene Generators, Stationary, Medium Pressure
409	Acetylene Generators, Stationary, Low Pressure
410	Slip Resistance of Floor Surface Materials
412	Refrigeration Unit Coolers
414	Meter Sockets (DoD)
416	Refrigerated Medical Equipment
427	Refrigerated Units (DoD)
429	Electrically Operated Valves (DoD)
430	Waste Disposers
437	Key Locks (DoD)
441	Gas Vents
443	Steel Auxiliary Tanks for Oil-Burner Fuel
444	Communications Cables (PE)
448	Pumps for Fire-Protection Service
452	Antenna Discharge Units
458	Power Converters/Inverters & Power-Converter/Inverter Systems for Land Vehicles and Marine Craft
462	Heat Reclaimers for Gas, Oil, or Solid Fuel-Fired Appliances
464	Audible Signal Appliances
465	Central Cooling Air Conditioners
466	Electric Scales
467	Grounding and Bonding Equipment
469	Musical Instruments and Accessories
471	Commercial Refrigerators and Freezers
474	Dehumidifiers
482	Portable Sun/heat Lamps
484	Room Air Conditioners
486A	Wire Connectors & Soldering Lugs for Use with Copper Conductors (DoD)
486B	Wire Connectors for Use with Aluminum Conductors
486C	Splicing Wire Connectors (DoD)
486D	Insulated Wire Connectors for Use with Underground Conductors
486E	Equipment Wiring Terminals for Use with Aluminum and/or Copper Conductors
489	Molded-Case Circuit Breakers and Circuit-Breaker Enclosures (DoD)
493	Thermoplastic-Insulated Underground Feeder & Branch-Circuit Cables
495	Power Operated Dispensing Devices for LP-Gas
496	Edison-Base Lampholders (DoD)
497	Protectors for Paired Conductor Communication Circuits
497A	Secondary Protectors for Communication Circuits
497B	Protectors for Data Communication and Fire Alarm Circuits

498	Attachment Plugs and Receptacles [DoD]
499	Electric Heating Appliances
506	Specialty Transformers [DoD]
507	Electric Fans [DoD]
508	Industrial Control Equipment
508C	Power Conversion Equipment
510	Electrical Insulated Tape
511	Porcelain Cleats, Knobs, and Tubes [DoD]
512	Fuseholders
514A	Metallic Outlet Boxes
514B	Fittings for Conduct and Outlet Boxes [DoD]
514C	Nonmetallic Outlet Boxes, Flush-Device Boxes, and Covers [DoD]
519	Impedance-Protected Motors
521	Heat Detectors for Fire Protective Signaling Systems
525	Flame Arresters for Use on Vents of Storage Tanks for Petroleum Oil & Gasoline [PE]
536	Flexible Metallic Hose
539	Single and Multiple Station Heat Detectors
541	Refrigerated Vending Machines
542	Lampholders, Starters, and Starter Holders for Fluorescent Lamps [DoD]
543	Impregnated-Fiber Electrical Conduit
544	Medical and Dental Equipment
547	Thermal Protectors for Motors
551	Transformer-Type Arc-welding Machines
555	Fire Dampers
555C	Ceiling Dampers
555S	Leakage Rated Dampers for Use in Smoke Control Systems
558	Industrial Trucks, Internal Combustion Engine Powered
559	Heat Pumps [DoD]
560	Electric Home Laundry Equipment
561	Floor Finishing Machines
563	Ice Makers
565	Liquid Level Gauges and Indicators for Anhydrous Ammonia & LP Gas
567	Pipe Connectors for Flammable and Combustible Liquids and LP Gas
569	Pigtails and Flexible Hose Connectors
574	Electric Oil Heaters
580	Tests for Uplift Resistance of Roof Assemblies [DoD]
583	Electric-Battery-Powered Industrial Trucks
586	High-Efficiency, Particulate, Air Filter Units [DoD]
588	Christmas Tree and Decorative Lighting Outfits
595	Marine-Type Electric Lighting Fixtures
603	Power Supplies for Use with Burglar-Alarm Systems
606	Linings and Screens for Use with Burglar Alarm Systems

608	Burglary Resistant Vault Doors and Modular Panels
609	Local Burglar Alarm Units & Systems
611	Central-Station Burglar Alarm Systems
618	Concrete Masonry Units
621	Ice Cream Makers
626	2-1/2 Gallon Stored-Pressure, Water Type Fire Extinguishers
632	Electrically Actuated Transmitters
634	Connectors and Switches for Use with Burglar Alarm Systems [DoD]
636	Holdup Alarm Units & Systems [DoD]
639	Intrusion Detection Units
641	Low Temperature Venting Systems, Type L
644	Container Assemblies for LP Gas
647	Unvented Kerosene-Fired Room Heaters and Portable Heaters
651	Schedule 40 and 80 Rigid PVC Conduit [DoD]
651A	Type EB and A Rigid PVC Conduit and HDPE Conduit [DoD]
664	Commercial Dry Cleaning Machines (Type IV)
668	Hose Valves for Fire Protection Service
674	Electric Motors and Generators for Use in Hazardous (Classified) Locations [DoD]
676	Underwater Lighting Fixtures
680	Emergency Vault Ventilators and Vault Ventilating Ports
681	Installation & Classification of Mercantile & Bank Burglar Alarm Systems [DoD, PE]
687	Burglary Resistant Safes
696	Electric Toys
697	Toy Transformers
698	Industrial Control Equipment for Use in Hazardous (Classified) Locations
705	Power Ventilators [DoD]
710	Exhaust Hoods for Commercial Cooking Equipment [DoD]
711	Fire Extinguishers, Rating and Fire Testing of [DoD]
719	Nonmetallic Sheathed Cables [DoD]
723	Test for Surface Burning Characteristics of Building Materials
726	Oil Fired Boiler Assemblies
727	Oil Fired Central Furnaces
729	Oil Fired Floor Furnaces
730	Oil Fired Wall Furnaces
731	Oil Fired Unit Heaters
732	Oil Fired Storage Tank Water Heaters
733	Oil Fired Air Heaters & Direct Fired Heaters
737	Fireplace Stoves
746A	Polymeric Materials - Short Term Property Evaluations
746B	Polymeric Materials - Long Term Property Evaluations [DoD]
746C	Polymeric Materials - Used in Electrical Equipment Evaluations
746D	Polymeric Materials - Fabricated Parts

746E	Polymeric Materials - Industrial Laminates, Filament Wound Tubing, Vulcanized Fiber, & Materials Used in Printing Wiring Boards [DoD]
749	Household Dishwashers
751	Vending Machines
752	Bullet Resisting Equipment
753	Alarm Accessories for Automatic Water Supply Control Valves for Fire Protection Service
756	Coin and Currency Changers and Actuators
763	Motor Operated Commercial Food Preparing Machines
768	Combination Locks [DoD, PE]
771	Night Depositories
773	Plug-In, Locking Type Photocontrols for Use in Area Lighting [DoD]
773A	Nonindustrial Photoelectric Switches for Lighting Control [DoD]
775	Graphic Arts Equipment
778	Motor Operated Water Pumps [DoD]
779	Electrically Conductive Floorings
781	Portable Electric Lighting Units for Use in Hazardous (Classified) Locations
783	Electric Flashlights and Lanterns for Use in Hazardous (classified) Locations
786	Key-Locked Safes (Class LK) [DoD]
789	Indicator Posts for Fire-Protection Service
790	Tests for Fire Resistance of Roof Covering Materials [DoD]
791	Residential Incinerators
795	Commercial Industrial Gas Heating Equipment [DoD]
796	Printed Wiring Boards
797	Electrical Metallic Tubing
810	Capacitors
813	Commercial Audio Equipment
814	Gas Tube Sign and Ignition Cable
817	Cord Sets and Power Supply Cords
823	Electric Heaters for Use in Hazardous (Classified) Locations
826	Household Electric Clocks
827	Central Stations for Watchman, Fire Alarm, and Supervisory Services
834	Heating, Water Supply and Power Boilers - Electric
840	Insulation Coordination Including Clearances and Creepage Distances for Electrical Equipment
842	Valves for Flammable Fluids
844	Electric Lighting Fixtures for Use in Hazardous (Classified) Locations
845	Motor Control Centers [DoD]
854	Service Entrance Cables
857	Busways and Associated Fittings [DoD]
858	Household Electric Ranges
858A	Safety Related Solid State Controls for Household Electric Ranges
859	Household Electric Personal Grooming Appliances

860	Pipe Unions for Flammable & Combustible Liquids & Fire Protection Services
863	Time Indicating and Recording Appliances
864	Control Units for Fire Protective Signaling Systems [DoD]
867	Electrostatic Air Cleaners [DoD]
869	Service Equipment [DoD]
869A	Reference Standard for Service Equipment
870	Wireways, Auxiliary Gutters, & Associated Fittings
873	Temperature Indicating and Regulating Equipment [DoD]
875	Electric Dry Bath Heaters
877	Circuit Breakers and Circuit Breaker Enclosures for Use in Hazardous (Classified) Locations
879	Electrode Receptacles for Gas Tube Signs
883	Fan Coil Units and Room Fan Heaters Units
884	Underfloor Raceways and Fittings [DoD]
886	Outlet Boxes and Fittings for Use in Hazardous (Classified) Locations [DoD]
887	Delayed Action Timelocks
891	Dead Front Switchboards [DoD]
894	Switches for Use in Hazardous (Classified) Locations
896	Oil Burning Stoves
900	Test Performances of Air Filter Units
907	Fireplace Accessories
910	Test for Flame Propagation and Smoke Density Valves for Electrical and Optical Fiber Cables Used in Spaces Transporting Environmental Air
912	Highway Emergency Signals
913	Intrinsically Safe Apparatus and Associated Apparatus for Use in Class I, II, and III, Division 1, Hazardous (Classified) Locations
916	Energy Management Equipment
917	Clock Operated Switches
921	Commercial Electric Dishwashers
923	Microwave Cooking Appliances [DoD]
924	Emergency Lighting and Power Equipment [DoD]
935	Fluorescent Lamp Ballasts
943	Ground Fault Circuit Interrupters
943A	Leakage Current Protection Devices
959	Medium Heat Appliances Factory Built Chimneys
961	Electric Hobby and Sports Equipment
964	Electrically Heated Bedding [PE]
969	Marking and Labeling Systems
972	Burglary Resisting Glazing Material
977	Fused Power Circuit Devices [DoD]
982	Motor Operated Household Food Preparing Machines
983	Surveillance Camera Units

984   Hermetic Refrigerant Motor Compressors [DoD]
985   Household Fire Warning System Units
987   Stationary and Fixed Electric Tools [PE]
991   Tests for Safety Related Controls Employing Solid State Devices
997   Wind Resistance of Prepared Roof Covering Materials [DoD]
998   Humidifiers
1002   Electrically Operated Valves for Use in Hazardous (Classified) Locations
1004   Electric Motors
1005   Electric Flatirons
1008   Automatic Transfer Switches [DoD]
1010   Receptacle Plug Combinations for Use in Hazardous (Classified) Locations
1012   Power Units Other than Class 2
1017   Vacuum Cleaners and Blower Cleaners
1018   Electric Aquarium Equipment
1020   Thermal Cutoffs for Use in Electrical Appliances and Components
1022   Line Isolation Monitors
1023   Household Burglar Alarm System Units
1025   Electric Air Heaters
1026   Electric Household Cooking and Food Serving Appliances
1028   Hair Clipping and Shaving Appliances
1029   High Intensity Discharge Lamp Ballasts [DoD]
1030   Sheathed Heating Elements [DoD]
1034   Burglary Resistant Electric Locking Mechanisms [DoD]
1037   Antitheft Alarms and Devices
1042   Electric Baseboard Heating Equipment [DoD]
1046   Grease Filters for Exhaust Ducts [PE]
1047   Isolated Power Systems Equipment
1053   Ground Fault Sensing and Relaying Equipment [DoD]
1054   Special Use Switches
1056   Fire Test of Upholstered Furniture
1058   Halogenated Agent Extinguishing System Units
1059   Terminal Blocks
1062   Unit Substations
1063   Machine-Tool Wires and Cables
1066   Low-Voltage AC and DC Power Circuit Breakers Used in Enclosures
1067   Electrically Conductive Equipment and Materials for Use in Flammable
       Anesthetizing Locations
1069   Hospital Signaling and Nurse Call Equipment
1072   Medium-Voltage Power Cables
1075   Gas Fired Cooking Appliances for Recreational Vehicles
1076   Proprietary Burglar Alarm Units and Systems [DoD]
1077   Supplementary Protectors for Use in Electrical Equipment

1081	Swimming Pool Pumps, Filters, and Chlorinates
1082	Household Electric Coffee Makers and Brewing-Type Appliances
1083	Household Electric Skillets and Frying-Type Appliances
1086	Household Trash Compactors
1087	Molded-Case Switches
1088	Temporary Lighting Strings
1090	Electric Snow Movers
1091	Butterfly Valves for Fire Protection Service
1093	Halogenated Agent Fire Extinguishers (DoD)
1096	Electric Central Air Heating Equipment (DoD)
1097	Double Insulation Systems for Use in Electrical Equipment
1100	Alcohol and Kerosene Cooking Appliances for Marine Use
1101	Solidified Fuel Cooking Appliances for Marine Use
1102	Nonintegral Marine Fuel Tanks
1104	Marine Navigation Lights
1105	Marine Use Filters, Strainers, and Separators
1106	Marine Manually Operated Shutoff Valves for Flammable Liquids
1110	Marine Combustible Gas Indicators
1111	Marine Carburetor Flame Arresters
1112	Marine Electric Motors and Generators (Cranking, Outdrive Tilt, Trim Tab, Generators, Alternators)
1113	Electrically Operated Pumps for Nonflammable Liquids, Marine
1114	Marine (USCG Type A) Flexible Fuel-Line Hose
1116	Marine Chain, Embarkation, and Pilot Ladders
1120	Flow-Through Marine Sanitation Devices
1121	Marine Engine Ignition Systems and Components
1123	Marine Through-Hull Fittings and Sea-Valves
1128	Marine Buoyant Devises
1129	Marine Blowers
1130	Wet Exhaust Components for Marine Engines
1133	Boat Circuit Breakers
1136	Marine Rigid and Flexible Air Ducting
1168	Recreational Boats
1175	Buoyant Cushions
1177	Buoyant Vests
1182	Marine Electric Outboard Propulsion Units (Trolling Motors)
1185	Portable Marine Fuel Tanks
1191	Components for Personal Flotation Devices
1193	Marine Filters and Strainers for Nonflammable Liquids
1196	Floating Waterlights (DoD)
1197	Immersion Suits
1198	Distress Signals

1199	Recreational Boats Less than 20 feet in Length
1203	Explosion-Proof and Dust-Ignition-Proof Electrical Equipment for Use in Hazardous (Classified) Locations
1206	Electric Commercial Clothes-Washing Equipment
1207	Sewage Plumps for Use in Hazardous (Classified) Locations
1230	Amateur Movie Lights
1236	Battery Chargers for Charging Engine Starter Batteries [DoD]
1238	Control Equipment for Use with Flammable Liquid Dispensing Devices [PE]
1240	Electric Commercial Clothes-Drying Equipment
1241	Junction Boxes for Swimming Pool Lighting Fixtures
1242	Intermediate Metal Conduit [DoD]
1244	Electrical and Electronic Measuring and Testing Equipment
1247	Diesel Engines for Driving Centrifugal Fire Pumps
1248	Engine-Generator Assemblies for Use in Recreational Vehicles
1254	Pre-Engineered Dry Chemical Extinguishing System Units
1256	Fire Rest of Roof Deck Constructions
1261	Electric Water Heaters for Pools and Tubs
1262	Laboratory Equipment
1270	Radio Receivers, Audio Systems, and Accessories
1275	Flammable Liquid Storage Cabinets
1277	Electrical Power and Control Tray Cables with Optional Optical Fiber Members
1278	Movable and Wall or Ceiling Hung Electric Room Heaters
1283	Electromagnetic Interference Filters
1285	Pipe and Couplings, Polyvinyl Chloride (PVC) for Underground Fire Service
1286	Office Furnishings
1296	Shear Resistance Tests for Ceiling Boards for Manufactured Homes
1298	Roof Trusses for Manufactured Homes
1310	Class 2 Power Units
1313	Nonmetallic Safety Cans for Petroleum Products
1314	Special Purpose Containers
1316	Glass, Fiber Reinforced Plastic Underground Storage Tanks for Petroleum Products, Alcohols, and Alcohol Gasoline Mixtures
1322	Fabricated Scaffold Planks and Stages
1323	Scaffold Hoists
1332	Organic Coatings for Steel Enclosures for Outdoor Use Electrical Equipment
1363	Temporary Power Taps
1409	Low-Voltage Video Products without Cathode Ray Tube Displays
1410	Television Receivers and High Voltage Video Products
1411	Transformers and Motor Transformers for Use in Audio, Radio, and Television Type Appliances
1412	Fusing Resistors and Temperature Limited Resistors for Radio and Television Type Appliances

1413	High Voltage Components for Television Type Appliances
1414	Across the Line, Antenna-Coupling, and Line by Pass Capacitors for Radio and Television Type Appliances
1416	Overcurrent and Overtemperature Protectors for Radio and Television Type Appliances
1417	Special Fuses for Radio and Television Type Appliances
1418	Cathode Ray Tubes
1419	Professional Video and Audio Equipment
1424	Cables for Power Limited Fire Protective Circuits
1426	Cables for Boats
1429	Pullout Switches
1431	Personal Hygiene and Health Care Appliances
1433	Control Centers for Changing Message Type Electric Signs
1436	Outlet Circuit Testers and Similar Indicating Devices
1437	Electrical Analog Instruments - Panel Board Types
1438	Household Electric Drip Type Coffee Makers
1439	Tests for Sharpness of Edges on Equipment
1441	Coated Electrical Sleeving
1445	Electric Water Bead Heaters
1446	Systems of Insulating Materials - General
1447	Electric Lawn Mowers
1448	Electric Hedge Trimmers
1449	Transient Voltage Surge Suppressors [PE]
1450	Motor-Operated Air Compressors, Vacuum Pumps, and Painting Equipment
1453	Electric Booster and Commercial Storage Tank Water Heaters
1459	Telephone Equipment
1468	Sprinkler Systems
1474	Adjustable Drop Nipples for Sprinkler Systems
1478	Fire Pump Relief Valves
1479	Fire Tests of Through Penetration Firestops
1480	Speakers for Fire Protective Signaling Systems
1481	Power Supplies for Fire Protective Signaling Systems
1482	Room Heaters, Solid Fuel Type
1484	Residential Gas Detectors
1486	Quick Opening Devices for Dry Pipe Valves for Fire Protection Service
1492	Audio-Video Products and Accessories
1500	Ignition-Protection Tests for Marine Products
1504	Oily Water Separating Equipment and Oil Content Meters
1506	Inflatable Boats
1517	Hybrid Personal Flotation Devices
1523	Controlled Descent Devices for Marine Use
1524	Carbon Monoxide Gas Detectors for Marine Use
1555	Electric Coin Operated Clothes Washing Equipment

1556   Electric Coin Operated Clothes Drying Equipment
1557   Electrically Isolated Semiconductor Devices
1558   Metal-Enclosed Low-Voltage Power Circuit Breaker Switchgear
1559   Insect-Control Equipment Electrocution Type [DoD]
1561   Dry Type General Purpose and Power Transformers
1562   Transformers, Distribution, Dry-Type - Over 600 volts
1563   Electric Spas, Equipment Assemblies, and Associated Equipment
1564   Industrial Battery Chargers
1565   Wire Positioning Devices
1567   Receptacles and Switches Intended for Use with Aluminum Wire
1569   Metal-Clad Cables [DoD]
1570   Fluorescent Lighting Fixtures [DoD]
1571   Incandescent Lighting Fixtures [DoD]
1572   High Intensity Discharge Lighting Fixtures [DoD]
1573   Stage and Studio Lighting Units
1574   Track Lighting Systems
1577   Optical Isolators
1581   Electrical Wires, Cables, and Flexible Cords
1585   Class 2 & Class 3 Transformers
1594   Sewing and Cutting Machines
1602   Gasoline-Engine-Powered, Rigid Cutting Member Edgers and Edger Trimmers
1604   Electrical Equipment for use in Class I & II, Division 2, and Class III
       Hazardous (Classified) Locations
1610   Central-Station Burglar Alarm Units
1624   Light Industrial Tools
1626   Residential Sprinklers for Fire Protection Service
1635   Digital Alarm Communicator System Units
1637   Home Health Care Signaling Equipment
1638   Visual Signaling Appliances
1641   Installation and Classification of Residential Burglar Alarm Systems
1642   Lithium Batteries
1647   Motor-Operated Massage and Exercise Machines
1651   Optical Fiber Cable
1659   Attachment Plug Blades for Use in Cord Sets and Power-Supply Cords
1660   Liquid Tight Flexible Nonmetallic Conduit
1662   Electric Chain Saws
1664   Immersion Detection Circuit Interrupters
1666   Test for Flame Propagation Height of Electrical and Optical Fiber
       Cables Installed Vertically in Shafts
1667   Tall Institutional Carts for use with Audio-Video, and Television Type equipment
1673   Electric Space Heating Cables

1676	Discharge Path Resistors
1678	Household, Commercial, and Professional Use Carts and Stands
1681	Wiring Device Configurations
1682	Plugs, Receptacles, and Cable Connectors, of the Pin and Sleeve Type
1684	Reinforced Thermosetting Resin Conduit
1685	Vertical-Tray Fire Propagation and Smoke Release Test for Electrical and Optical Fiber Cables
1686	Pin and Sleeve Configurations
1690	Data-Processing Cable
1703	Flat Plate Photovoltaic Modules and Panels
1709	Rapid Rise Fire Tests of Protection Materials for Structural Steel
1711	Amplifiers for Fire Protective Signaling Systems
1713	Pressure Pipe and Couplings, Glass Fiber-Reinforced, for Underground Fire Service
1715	Fire Test of Interior Finish Material
1726	Automatic Drain Valves for Standpipe Systems
1727	Commercial Electric Personal Grooming Appliances
1730	Smoke Detector Monitors and Accessories for Individual Living Units or Multifamily Residences and Hotel/Motel Rooms
1738	Venting Systems for Gas Burning Appliances, Categories II, III, and IV
1739	Pilot-Operated Pressure Control Valves for Fire Protection Service
1746	External Corrosion Protection Systems for Steel Underground Storage Tanks
1767	Early-Suppression Fast-Response Sprinklers
1769	Cylinder Valves
1773	Termination Boxes
1776	High-pressure Cleaning Machines
1777	Chimney Liners
1778	Uninterruptible Power Supply Equipment
1784	Air Leakage Tests for Door Assemblies
1786	Nightlights
1795	Hydromassage Bathtubs
1803	Factory Follow-Up on Third Party Certified Portable Fire Extinguishers
1812	Ducted Heat Recovery Ventilators
1815	Nonducted Heat Recovery Ventilators
1820	Fire Tests of Pneumatic Tubing for Flame and Smoke Characteristics
1853	Nonreusable Plastic Containers for Flammable and Combustible Liquids
1863	Communication Circuit Accessories
1876	Isolating Signal and Feedback Transformers for Use in Electronic Equipment
1887	Fire Test of Plastic Sprinkler Pipe for Flame and Smoke Characteristics
1895	Fire Tests of Mattresses
1897	Uplift Tests for Roof Covering Systems
1917	Solid-State Fan Speed Controls

1950   Safety of Information Technology Equipment, Including Electrical Business Equipment
1963   Refrigerant Recovery/Recycling Equipment
1971   Signaling Devices for the Hearing Impaired
1975   Fire Tests for Foamed Plastics Used for Decorative Purposes
1989   Standby Batteries
1993   Self-Ballasted Lamps and Lamp Adapters
1994   Low Level Path Marking and Lighting Systems
1995   Heating and Cooling Equipment (DoD)
1996   Duct Heaters
1998   Safety-Related Software
2006   Halon 1211 Recovery/Recharge Equipment
2021   Fixed and Location-Dedicated Electric Room Heaters
2034   Single and Multiple Station Carbon Monoxide Detectors
2043   Fire Test for Heat and Visible Smoke Release for Discrete Products and Their Accessories Installed in Air-Handling Spaces
2044   Commercial Closed Circuit Television Equipment
2050   Defense Industrial Security Systems for the Protection of Classified Materials
2083   Halon 1301 Recovery/Recycling Equipment
2157   Electric Clothes Washing Machines and Extractors
2158   Electric Clothes Dryers
2170   Field Conversion/Retrofit of Products to Change to an Alternative Refrigerant – Construction and Operation
2171   Field Conversion/Retrofit of Products to Change to an Alternative Refrigerant – Insulating Material and Refrigerant Compatibility
2172   Field Conversion/Retrofit of Products to Change to an Alternative Refrigerant – Procedures and Methods
2096   Commercial Industrial Gas and/or Oil Burning Assemblies with Emission Reduction Equipment
2097   Reference Standard for Double Insulation Systems for Use in Electronic Equipment
3101-1 Electrical Equipment for Laboratory Use; Part 1: General Requirements
8730-1 Electrical Controls for Household and Similar Use; Part 1: General Requirements (DoD)

Appendix B

Underwriters Laboratories

Publications in Alphabetical Order

NOTE: (DoD) = Department of Defense
(PE) = Professional Engineer

626	2-1/2 Gallon Stored-Pressure, Water Type Fire Extinguishers
294	Access Control System Units
297	Acetylene Generators Portable, Medium-Pressure
409	Acetylene Generators, Stationary, Low Pressure
408	Acetylene Generators, Stationary, Medium Pressure
1474	Adjustable Drop Nipples for Sprinkler Systems
1784	Air Leakage Tests for Door Assemblies
193	Alarm Valves for Fire-Protection Service
1037	Alarms and Devices, Antitheft
1100	Alcohol and Kerosene Cooking Appliances for Marine Use
1230	Amateur Movie Lights
1711	Amplifiers for Fire Protective Signaling Systems
22	Amusement and Gaming Machines
452	Antenna Discharge Units
150	Antenna Rotators
1414	Antenna-Coupling, and Line by Pass Capacitors for Radio and Television Type Appliances, Across the Line
4	Armored Cable (DoD)
107	Asbestos-Cement Pipe and Couplings
498	Attachment Plugs and Receptacles (DoD)
464	Audible Signal Appliances
1492	Audio-Video Products and Accessories
291	Automated Teller Systems
1726	Automatic Drain Valves for Standpipe Systems
199	Automatic Sprinklers for Fire-Protection Service
1008	Automatic Transfer Switches (DoD)
395	Automotive Fuel Tanks
275	Automotive Glass-Tube Fuses
1236	Battery Chargers for Charging Engine Starter Batteries (DoD)
1133	Boat Circuit Breakers

1177	Buoyant Vests
752	Bullet Resisting Equipment
1175	Buoyant Cushions
1034	Burglary Resistant Electric Locking Mechanisms (DoD)
687	Burglary Resistant Safes
908	Burglary Resistant Vault Doors and Modular Panels
972	Burglary Resisting Glazing Material
857	Busways and Associated Fittings (DoD)
1426	Cables for Boats
1424	Cables for Power Limited Fire Protective Circuits
810	Capacitors
1524	Carbon Monoxide Gas Detectors for Marine Use
154	Carbon Dioxide Fire Extinguishers
1418	Cathode Ray Tubes
555C	Ceiling Dampers
209	Cellular Metal Floor Raceways and Fittings (DoD)
465	Central Cooling Air Conditioners
827	Central Stations for Watchman, Fire Alarm, and Supervisory Services
611	Central-Station Burglar Alarm Systems
1610	Central-Station Burglar Alarm Units
218A	Centrigugal Fire Pumps, Battery Contractors for Use in Diesel Engines Driving
312	Check Valves for Fire-Protection Services
1777	Chimney Liners
103	Chimneys, Factor-Built Residential Type & Building Heating Appliances (DoD)
588	Christmas Tree and Decorative Lighting Outfits
877	Circuit Breakers and Circuit Breaker Enclosures for Use in Hazardous (Classified) Locations
1585	Class 2 & Class 3 Transformers
1310	Class 2 Power Units
198B	Class H Fuses
198D	Class K Fuses
198E	Class R Fuses (DoD)
198H	Class T Fuses (DoD)
917	Clock Operated Switches
181A	Closure Systems for Use with Rigid Air Ducts & Air Connectors
1441	Coated Electrical Sleeving
756	Coin and Currency Changers and Actuators
768	Combinaition Locks (DoD, PE)
813	Commercial Audio Equipment
2044	Commercial Closed Circuit Television Equipment
664	Commercial Dry Cleaning Machines (Type IV)

197   Commercial Electric Cooking Appliances
921   Commercial Electric Dishwashers
1727   Commercial Electric Personal Grooming Appliances
795   Commercial Indsutrial Gas Heating Equipment [DoD]
2096   Commercial Industrial Gas and/or Oil Burning Assemblies
       with Emission Reduction Equipment
471   Commercial Refrigerators and Freezers
1863   Communication Circuit Accessories
444   Communications Cables [PE]
252A   Compressed Gas Regulator Accessories
1191   Components for Personal Flotation Devices
252   Compressed Gas Regulators [DoD]
618   Concrete Masonry Units
634   Connectors and Switches for use with Burglar Alarm Systems [DoD]
352   Constant-Level Oil Valves
644   Container Assemblies for LP Gas
1433   Control Centers for Changing Message Type Electric Signs
1238   Control Equipment for Use with Flammable Liquid Dispensing Devices [PE]
864   Control Units for Fire Protective Signaling Systems [DoD]
1523   Controlled Descent Devices for Marine Use
355   Cord Reels
817   Cord Sets and Power Supply Cords
1659   Cord Sets and Power-Supply Cords, Attachment Plug Blades for Use in
1769   Cylinder Valves
198L   D-C Fuses for Industrial Use
1690   Data-Processing Cable
891   Dead Front Switchboards [DoD]
2050   Defense Industrial Security Systems for the Protection of Classified Materials
474   Dehumidifiers
887   Delayed Action Timelocks
1247   Diesel Engines for Driving Centrifugal Fire Pumps
1635   Digital Alarm Communicator System Units
1676   Discharge Path Resistors
1198   Distress Signals
228   Door Closers_Holders, with or without Intergral Smoke Detectors
325   Door, Drapery, Gate, Louver, & Window Operators and Systems [DoD]
1097   Double Insulation Systems for use in Electrical Equipment
378   Draft Equipment
399   Drinking-Water Coolers [PE]
299   Dry Chemical Fire Extinguishers [DoD]
260   Dry Pipe & Deluge Valves for Fire-Protection Service

1561	Dry Type General Purpose and Power Transformers
1996	Duct Heaters
1812	Ducted Heat Recovery Ventilators
1767	Early-Suppression Fast-Response Sprinklers
496	Edison-Base Lampholders (DoD)
1018	Electric Aquarium Equipment
510	Elecrical Insulated Tape
1004	Electric Motors
1025	Electric Air Heaters
1042	Electric Baseboard Heating Equipment (DoD)
1453	Electric Booster and Commercial Storage Tank Water Heaters
1096	Electric Central Air Heating Equipment (DoD)
1662	Electric Chain Saws
2158	Electric Clothes Dryers
2157	Electric Clothes Washing Machines and Extractors
1556	Electric Coin Operated Clothes Drying Equipment
1555	Electric Coin Operated Clothes Washing Equipment
1240	Electric Commercial Clothes-Drying Equipment
1206	Electric Commercial Clothes-Washing Equipment
875	Electric Dry Bath Heaters
507	Electric Fans (DoD)
783	Electric Flashlights and Lanterns for Use in Hazardous (Classified) Locations
1005	Electric Flatirons
82	Electric Gardening Appliances
823	Electric Heaters for Use in Hazardous (Classified) Locations
499	Electric Heating Appliances
130	Electric Heating Pads
1448	Electric Hedge Trimmers
961	Electric Hobby and Sports Equipment
560	Electric Home Laundry Equipment
1026	Electric Household Cooking and Food Serving Appliances
1447	Electric Lawn Mowers
844	Electric Lighting Fixtures for Use in Hazardous (Classified) Locations
674	Electric Motors and Generators for Use in Hazardous (Classified) Locations (DoD)
574	Electric Oil Heaters
466	Electric Scales
48	Electric Signs
1090	Electric Snow Movers
1673	Electric Space Heating Cables
1563	Electric Spas, Equipment Assemblies, and Associated Equipment
696	Electric Toys

1445	Electric Water Bead Heaters
1261	Electric Water Heaters for Pools and Tubs
583	Electric-Battery-Powered Industrial Trucks
69	Electric-Fence Controllers
1437	Electrical Analog Instruments - Panel Board Types
1244	Electrical and Electronic Measuring and Testing Equipment
8730-1	Electrical Controls for Household and Similar Use; Part 1: General Requirements (DoD)
3101-1	Electrical Equipment for Laboratory Use; Part 1: General Requirements
1604	Electrical Equipment for Use in Class I & II, Division 2, and Class III Hazardous (Classified) Locations
797	Electrical Metallic Tubing
1277	Electrical Power and Control Tray Cables with Optional Optical Fiber Members
310	Electrical Quick-Connect Terminals
1581	Electrical Wires, Cables, and Flexible Cords
632	Electrically Actuated Transmitters
1067	Electrically Conductive Equipment and Materials for Use in Flammable Anesthetizing Locations
779	Electrically Conductive Floorings
964	Electrically Heated Bedding (PE)
1557	Electrically Isolated Semiconductor Devices
1113	Electrically Operated Pumps for Nonflammable Liquids, Marine
429	Electrically Operated Valves (DoD)
1002	Electrically Operated Valves for Use in Hazardous (Classified) Locations
879	Electrode Receptacles for Gas Tube Signs
1283	Electromagnetic Interference Filters
867	Electrostatic Air Cleaners (DoD)
104	Elevator Door Locking Devices and Contacts
924	Emergency Lighting and Power Equipment (DoD)
680	Emergency Vault Ventilators and Vault Ventilating Ports
98	Enclosed and Dead-Front Switches (DoD)
50	Enclosures for Electrical Equipment
916	Energy Management Equipment
1248	Engine-Generator Assemblies for Use in Recreational Vehicles
486E	Equipment Wiring Terminals for Use with Aluminum and/or Copper Conductors
710	Exhaust Hoods for Commercial Cooking Equipment (DoD)
1203	Explosion-Proof and Dust-Ignition-Proof Electrical Equipment for Use in Hazardous (Classified) Locations
1746	External Corrosion Protection Systems for Steel Underground Storage Tanks
224	Extruded Insulated Tubing
1322	Fabricated Scaffold Planks and Stages
1803	Factory Follow-Up on Third Party Certified Portable Fire Extinguishers

127	Factory-Build Fireplaces
181	Factory-Made Air Ducts and Air Connectors
883	Fan Coil Units and Room Fan Heaters Units
2170	Field Conversion/Retrofit of Products to Change to an Alternative Refrigerant - Construction and Operation
2171	Field Conversion/Retrofit of Products to Change to an Alternative Refrigerant - Insulating Material and Refrigerant Compatibility
2172	Field Conversion/Retrofit of Products to Change to an Alternative Refrigerant - Procedures and Methods
555	Fire Dampers
405	Fire Department Connections
92	Fire Extinguisher and Booster Hose
711	Fire Extinguishers, Rating and Fire Testing of [DoD]
753	Fire Protection Service, Alarm Accessories for Automatic Water Supply Control Valves for
1091	Fire Protection Service, Butterfly Valves for
1478	Fire Pump Relief Valves
1256	Fire Test of Roof Deck Constructions
2043	Fire Test for Heat and Visible Smoke Release for Discrete Products and Their Accessories Installed in Air-Handling Spaces
1887	Fire Test of Plastic Sprinkler Pipe for Flame and Smoke Characteristics
1715	Fire Test of Interior Finish Material
1056	Fire Test of Upholstered Furniture
300	Fire Testing/ Fire Extinguishing Systems for Protection of Restaurant Cooking Areas
1975	Fire Tests for Foamed Plastics Used for Decorative Purposes
263	Fire Tests of Building Construction and Materials
10B	Fire Tests of Door Assemblies
1895	Fire Tests of Mattresses
1820	Fire Tests of Pneumatic Tubing for Flame and Smoke Characteristics
1479	Fire Tests of Through Penetration Firestops
9	Fire Tests of Window Assemblies
907	Fireplace Accessories
737	Fireplace Stoves
514B	Fittings for Conduit and Outlet Boxes [DoD]
2021	Fixed and Location-Dedicated Electric Room Heaters
525	Flame Arresters for Use on Vents of Storage Tanks for Petroleum Oil & Gasoline [PE]
1275	Flammable Liquid Storage Cabinets
1703	Flat Plate Photovoltaic Modules and Panels
62	Flexible Cord and Fixture Wire [DoD]
1	Flexible Metal Conduit
536	Flexible Metallic Hose

3	Flexible Nonmetallic Tubing for Electrical Wiring[DoD]
1196	Floating Waterlights [DoD]
561	Floor Finishing Machines
1120	Flow-Through Marine Sanitation Devices
935	Fluorescent Lamp Ballasts
1570	Fluorescent Lighting Fixtures [DoD]
162	Foam Equipment and Liquid Concentrates
8	Foam Fire Extinguishers
977	Fused Power Circuit Devices [DoD]
512	Fuseholders
198G	Fuses for Supplementary Overcurrent Protection
1412	Fusing Resistors and Temperature Limited Resistors for Radio and Television Type Appliances
141	Garment Finishing Appliances
307B	Gas Burning Heating Appliances for Mobile Homes & Recreational Vehicles
1075	Gas Fired Cooking Appliances for Recreational Vehicles
814	Gas Tube Sign and Ignition Cable
441	Gas Vents
194	Gasketed Joints for Ductile-Iron Pipe & Fittings for Fire Protection Service
157	Gaskets and Seals
1602	Gasoline-Engine-Powered, Rigid Cutting Member Edgers and Edger Trimmers
262	Gate Valves for Fire-Protection Service [DoD]
404	Gauges, Indicating Pressure, for Compressed Gas Service
20	General-Use Snap Switches [DoD]
1316	Glass, Fiber Reinforced Plastic Underground Storage Tanks for Petroleum Products, Alcohols, and Alcohol Gasoline Mixtures
775	Graphic Arts Equipment
1046	Grease Filters for Exhaust Dusts [PE]
943	Ground Fault Circuit Interrupters
1053	Ground Fault Sensing and Relaying Equipment [DoD]
467	Grounding and Bonding Equipment
1028	Hair Clipping and Shaving Appliances
1058	Halogenated Agent Extinguishing System Units
1093	Halogenated Agent Fire Extinguishers [DoD]
2006	Halon 1211 Recovery/Recharge Equipment
2083	Halon 1301 Recovery/Recycling Equipment
147	Hand-Held Torches for Fuel Gases
521	Heat Detectors for Fire Protective Signaling Systems
559	Heat Pumps [DoD]
462	Heat Reclaimers for Gas, Oil, or Solid Fuel-Fired Appliances
33	Heat Responsive Links for Fire-Protection Service

1995	Heating and Cooling Equipment (DoD)
834	Heating, Water Supply and Power Boilers - Electric
984	Hermetic Refrigerant Motor Compressors (DoD)
1029	High Intensity Discharge Lamp Ballasts (DoD)
1572	High Intensity Discharge Lighting Fixtures (DoD)
1413	High Voltage Components for Television Type Appliances
347	High Voltage Industrial Control Equipment
586	High-Efficiency, Particulated, Air Filter Units (DoD)
198C	High-Interrupting-Capacity Fuses, Current-Limited Fuses (DoD)
1776	High-pressure Cleaning Machines
912	Highway Emergency Signals
636	Holdup Alarm Units & Systems (DoD)
1637	Home Health Care Signaling Equipment
330	Hose & Hose Assemblies for Dispensing Gasoline (PE)
668	Hose Valves for Fire Protection Service
250	Household Refrigerators and Freezers
749	Household Dishwashers
1069	Hospital Signaling and Nurse Call Equipment
1023	Household Burglar Alarm System Units
826	Household Electric Clocks
1082	Household Electric Coffee Makers and Brewing-Type Appliances
1438	Household Electric Drip Type Coffee Makers
859	Household Electric Personal Grooming Appliances
858	Household Electric Ranges
1083	Household Electric Skillets and Frying-Type Appliances
174	Household Electric Storage Tank Water Heaters
985	Household Fire Warning System Units
1086	Household Trash Compactors
1678	Household, Commercial, and Professional Use Carts and Stands
998	Humidifiers
1517	Hybrid Personal Flotation Devices
246	Hydrants for Fire-Protection Service
1795	Hydromassage Bathtubs
621	Ice Cream Makers
563	Ice Makers
1500	Ignition-Protection Tests for Marine Products
1664	Immersion Detection Circuit Interrupters
1197	Immersion Suits
519	Impedance-Protected Motors
543	Impregnated-Fiber Electrical Conduit
1571	Incandescent Lighting Fixtures (DoD)

393	Indicating Pressure Gauges for Fire Protection Services
789	Indicator Posts for Fire-Protection Service
1564	Industrial Battery Chargers
508	Industrial Control Equipment
698	Industrial Control Equipment for Use in Hazardous (Classified) Locations
558	Industrial Trucks, Internal Combustion Engine Powered
1506	Inflatable Boats
1559	Insect-Control Equipment Electrocution Type [DoD]
681	Installation & Classification of Mercantile & Bank Burglar Alarm Systems [DoD, PE]
1641	Installation and Classification of Residential Burglar Alarm Systems
96A	Installation Requirements for Lightning Protection Systems
486D	Insulated Wire Connectors for Use with Underground Conductors
840	Insulation Coordination Including Clearances and Creepage Distances for Electrical Equipment
1242	Intermediate Metal Conduit [DoD]
913	Intrinsically Safe Apparatus and Associated Apparatus for Use in Class I, II, and III, Division 1, Hazardous (Classified) Locations
639	Intrusion Detection Units
1047	Isolated Power Systems Equipment
1876	Isolating Signal and Feedback Transformers for Use in Electronic Equipment
1241	Junction Boxes for Swimming Pool Lighting Fixtures
437	Key Locks [DoD]
786	Key-Locked Safes (Class LK) [DoD]
363	Knife Switch
1262	Laboratory Equipment
542	Lampholders, Starters, and Starter Holders for Fluorescent Lamps [DoD]
943A	Leakage Current Protection Devices
555S	Leakage Rated Dampers for Use in Smoke Control Systems
1624	Light Industrial Tools
96	Lighting Protection Components
1022	Line Isolation Monitors
19	Lined Fire Hose and Hose Assemblies [DoD]
219	Lined Fire Hose for Interior Standpipes
606	Linings and Screens for Use with Burglar Alarm Systems
307A	Liquid Fuel-Burning Heating Appliances for Manufactured Homes & Rec. Vehicles
565	Liquid Level Gauges and Indicators for Anhydrous Ammonia & LP Gas
1660	Liquid Tight Flexible Nonmetallic Conduit
180	Liquid-Level Indicating Gauges & Tank-Filling Signals for Petroleum Products
353	Liquid-Tight Flexible Steel Conduit [DoD]
360	Liquid-Tight Flexible Steel Conduit [DoD]
1642	Lithium Batteries

609	Local Burglar Alarm Units & Systems
1994	Low Level Path Marking and Lighting Systems
641	Low Temperature Venting Systems, Type L
1066	Low-Voltage AC and DC Power Circuit Breakers used in Enclosures
1409	Low-Voltage Video Products without Cathode Ray Tube Displays
234	Low Voltage Lighting Fixtures for Use in Recreational Vehicles
21	LP-Gas Hose
1063	Machine-Tool Wires and Cables
407	Manifolds for Compressed Gases
38	Manually Actuated Signaling Boxes for use with Fire-Protective Signaling Systems[DoD]
183	Manufactured Wiring Systems
1114	Marine (USCG Type A) Flexible Fuel-Line Hose
1129	Marine Blowers
1128	Marine Buoyant Devises
1111	Marine Carburetor Flame Arresters
1116	Marine Chain, Embarkation, and Pilot Ladders
1110	Marine Combustible Gas Indicators
1112	Marine Electric Motors and Generators (Cranking, Outdrive Tilt, Trim Tab, Generators, Alternators)
1182	Marine Electric Outboard Propulsion Units (Trolling Motors)
1121	Marine Engine Ignition Systems and Components
1193	Marine Filters and Strainers for Nonflammable Liquids
1106	Marine Manually Operated Shutoff Valves for Flammable Liquids
1104	Marine Navigation Lights
1136	Marine Rigid and Flexible Air Ducting
1123	Marine Through-Hull Fittings and Sea-Valves
1105	Marine Use Filters, Strainers, and Separators
595	Marine-Type Electric Lighting Fixtures
969	Marking and Labeling Systems
55A	Materials for Built-Up Roof Coverings
544	Medical and Dental Equipment
959	Medium Heat Appliances Factory Built Chimneys
1072	Medium-Voltage Power Cables
30	Metal Safety Cans
32	Metal Waste Cans
1569	Metal-Clad Cables [DoD]
1558	Metal-Enclosed Low-Voltage Power Circuit Breaker Switchgear
514A	Metallic Outlet Boxes
414	Meter Sockets [DoD]
25	Meters for Flammable & combustible Liquids and LP-Gas

923	Microwave Cooking Appliances [DoD]
198M	Mine-Duty Fuses
489	Molded-Case Circuit Breakers and Circuit-Breaker Enclosures [DoD]
1087	Molded-Case Switches
845	Motor Control Centers [DoD]
763	Motor Operated Commercial Food Preparing Machines
982	Motor Operated Household Food Preparing Machines
778	Motor Operated Water Pumps [DoD]
1450	Motor-Operated Air Compressors, Vacuum Pumps, and Painting Equipment
73	Motor-Operated Appliances
1647	Motor-Operated Massage and Exercise Machines
1278	Movable and Wall or Ceiling Hung Electric Room Heaters
469	Musical Instruments and Accessories
771	Night Depositories
1786	Nightlights
1815	Nonducted Heat Recovery Ventilators
773A	Nonindustrial Photoelectric Switches for Lighting Control [DoD]
1102	Nonintegral Marine Fuel Tanks
242	Nonmetallic Containers for Waste Paper
514C	Nonmetallic Outlet Boxes, Flush-Device Boxes, and Covers [DoD]
1313	Nonmetallic Safety Cans for Petroleum Products
719	Nonmetallic Sheathed Cables [DoD]
147B	Nonrefillable (Disposable) Type Metal Container Assemblies for Butane
147A	Nonrefillable (Disposable) Type Fuel Gas Cylinder Assemblies
1853	Nonreusable Plastic Containers for Flammable and Combustible Liquids
1286	Office Furnishings
296	Oil Burners [DoD]
896	Oil Burning Stoves
733	Oil Fired Air Heaters & Direct Fired Heaters
726	Oil Fired Boiler Assemblies
727	Oil Fired Central Furnaces
729	Oil Fired Floor Furnaces
732	Oil Fired Storage Tank Water Heaters
731	Oil Fired Unit Heaters
730	Oil Fired Wall Furnaces
1504	Oily Water Separating Equipment and Oil Content Meters
1651	Optical Fiber Cable
1577	Optical Isolators
1332	Organic Coatings for Steel Enclosures for Outdoor use Electrical Equipment
886	Outlet Boxes and Fittings for use in Hazardous (Classified) Locations [DoD]
1436	Outlet Circuit Testers and Similar Indicating Devices

1416	Overcurrent and Overtemperature Protectors for Radio and Television Type Appliances
123	Oxy-Fuel Gas Torches
67	Panelborads
305	Panic Hardware
1431	Personal Hygiene and Health Care Appliances
122	Photographic Equipment
569	Pigtails and Flexible Hose Connectors
1739	Pilot-Operated Pressure Control Valves for Fire Protection Service
1686	Pin and Sleeve Configurations
860	Pipe Unions for Flammable & Combustible Liquids & Fire Protection Services
1285	Pipe and Couplings, Polyvinyl Chloride (PVC) for Underground Fire Service
567	Pipe Connectors for Flammable and Combustible Liquids and LP Gas
203	Pipe Hanger Equipment for Fire-Protection Service
385	Play Pipes for Water Supply Testing in Fire-Protection Services
198F	Plug Fuses
773	Plug-In, Locking Type Photocontrols for use in Area Lighting[DoD]
1682	Plugs, Receptacles, and Cable Connectors, of the Pin and Sleeve Type
365	Police Station Connected Burglar Alarm Units & Systems
746D	Polymeric Materials - Fabricated Parts
746E	Polymeric Materials - Industrial Laminates, Filament Wound Tubing, Vulcanized Fiber, & Materials Used in Printing Wiring Boards [DoD]
746B	Polymeric Materials - Long Term Property Evaluations [DoD]
746A	Polymeric Materials - Short Term Property Evaluations
746C	Polymeric Materials - Used in Electrical Equipment Evaluations
511	Porcelain Cleats, Knobs, and Tubes [DoD]
153	Portable Electric Lamps
781	Portable Electric Lighting Units for use in Hazardous (Classified) Locations
45	Portable Electric Tools [DoD]
1185	Portable Marine Fuel Tanks
184	Portable Metal Ladders
401	Portable Spray Hose Nozzles for Fire Protection Service
482	Portable Sun/heat Lamps
112	Portable Wood Ladders
298	Portable Electric Hand Lamps [DoD]
351	Posettes
508C	Power Conversion Equipment
458	Power Converters//Inverters & Power-Converter/Inverter Systems for Land Vehicles and Marine Craft
495	Power Operated Dispensing Devices for LP-Gas
231	Power Outlets

1481	Power Supplies for Fire Protective Signaling Systems
603	Power Supplies for Use with Burglar-Alarm Systems
1012	Power Units other than Class 2
705	Power Ventilators (DoD)
13	Power-Limited Circuit Cables
87	Power-Operated Dispensing Devices for Petroleum Products (DoD)
51	Power-Operated Pumps for Anhydrous Ammonia and LP-Gas
79	Power-Operated Pumps for Petroleum Product Dispensing Systems
1254	Pre-Engineered Dry Chemical Extinguishing System Units
136	Pressure Cookers
1713	Pressure Pipe and Couplings, Glass Fiber-Reinforced, for Underground Fire Service
144	Pressure Regulating Valves for LP-Gas
372	Primary Safety Controls for Gas & Oil-Fired Appliances
796	Printed Wiring Boards
1419	Professional Video and Audio Equipment
1076	Proprietary Burglar Alarm Units and Systems (DoD)
497B	Protectors for Data Communication and Fire Alarm Circuits
497	Protectors for Paired Conductor Communication Circuits
1429	Pullout Switches
448	Pumps for Fire-Protection Service
343	Pumps for Oil-Burning Appliances
1486	Quick Opening Devices for Dry Pipe Valves for Fire Protection Service
1270	Radio Receivers, Audio Systems, and Accessories
1709	Rapid Rise Fire Tests of Protection Materials for Structural Steel
1010	Receptacle Plug Combinations for use in Hazardous (Classified) Locations
1567	Receptacles and Switches Intended for Use with Aluminum Wire
1168	Recreational Boats
1199	Recreational Boats Less than 20 feet in Length
2097	Reference Standard for Double Insulation Systems for use in Electronic Equipment
869A	Reference Standard for Service Equipment
1963	Refrigerant Recovery/Recycling Equipment
207	Refrigerant-Containing Components & Accessories, Nonelectric (PE)
416	Refrigerated Medical Equipment
427	Refrigerated Units (DoD)
541	Refrigerated Vending Machines
303	Refrigeration and Air-Conditioning Condensing and Compressor Units
412	Refrigeration Unit Coolers
1684	Reinforced Thermosetting Resin Conduit
140	Relocking Devices for Safes and Vaults
1484	Residential Gas Detectors
791	Residential Incinerators

1626	Residential Sprinklers for Fire Protection Service
6	Rigid Metal Conduit
213	Rubber Gasketed Fittings for Fire Protection Service
311	Roof Jacks for Manufactured Homes and Recreational Vehicles
1298	Roof Trusses for Manufactured Homes
484	Room Air-Conditioners
1482	Room Heaters, Solid Fuel Type
44	Rubber-Insulated Wires and Cables
1950	Safety of Information Technology Equipment, Including Electrical Business Equipment
858A	Safety Related Solid State Controls for Household Electric Ranges
132	Safety Relief Valves for Anhydrous Ammonia & LP-Gas
1998	Safety-Related Software
1323	Scaffold Hoists
651	Schedule 40 and 80 Rigid PVC Conduit (DoD)
497A	Secondary Protectors for Communication Circuits
1993	Self-Ballasted Lamps and Lamp Adapters
47	Semiautomatic Fire Hose Storage Devices
70	Septic Tanks, Situminous Coated Metal
854	Service Entrance Cables
869	Service Equipment (DoD)
1207	Sewage Pumps for use in Hazardous (Classified) Locations
1594	Sewing and Cutting Machines
1296	Shear Resistance Tests for Ceiling Boards for Manufactured Homes
1030	Sheathed Heating Elements (DoD)
1971	Signaling Devices for the Hearing Impaired
2034	Single and Multiple Station Carbon Monoxide Detectors
539	Single and Multiple Station Heat Detectors
217	Single and Multiple Station Smoke Detectors
14B	Sliding Hardware for Standard, Horizontally Mounted Tin-Clad Fire Doors
410	Slip Resistance of Floor Surface Materials
1730	Smoke Detector Monitors and Accessories for Individual Living Units or Multifamily Residences and Hotel/Motel Rooms
268A	Smoke Detectors For Duct Application
268	Smoke Detectors for Fire Protective Signaling Systems
391	Solid-Fuel & Combination-Fuel Central & Supplementary Furnaces
244A	Solid-State Controls for Appliances
1917	Solid-State Fan Speed Controls
1101	Solidified Fuel Cooking Appliances for Marine Use
1480	Speakers for Fire Protective Signaling Systems
1417	Special Fuses for Radio and Television Type Appliances
1314	Special Purpose Containers

1054	Special Use Switches
506	Specialty Transformers (DoD)
486C	Splicing Wire Connectors (DoD)
1468	Sprinkler Systems
1573	Stage and Studio Lighting Units
1989	Standby Batteries
987	Stationary and Fixed Electric Tools (PE)
142	Steel Aboveground Tanks for Flammable & Combustible Liquids
443	Steel Auxiliary Tanks for Oil-Burner Fuel
80	Steel Inside Tanks for Oil-Burner Fuel (DoD)
58	Steel Underground Tanks for Flammable & Combustible Liquids (DoD)
331	Strainers for Flammable Fluids and Anhydrous Ammonia
1081	Swimming Pool Pumps, Filters, and Chlorinators
1077	Supplementary Protectors for use in Electrical Equipment
5	Surface Metal Raceways and Fittings (DoD, PE)
983	Surveillance Camera Units
14C	Swinging Hardware for Standard Tin-Clad Fire Doors Mounted Singly & in Pairs
894	Switches for use in Hazardous (Classified) Locations
1446	Systems of Insulating Materials - General
1667	Tall Institutional Carts for use with Audio-Video, and Television Type equipment
1459	Telephone Equipment
1410	Television Receivers and High Voltage Video Products
873	Temperature Indicating and Regulating Equipment (DoD)
1088	Temporary Lighting Strings
1363	Temporary Power Taps
1059	Terminal Blocks
1773	Termination Boxes
493	Thermoplastic-Insulated Underground Feeder & Branch-Circuit Cables
910	Test for Flame Propagation and Smoke Density Valves for Electrical and Optical Fiber Cables used in Spaces Transporting Environmental Air
1666	Test for Flame Propagation Height of Electrical and Optical Fiber Cables installed Vertically in Shafts
723	Test for Surface Burning Characteristics of Building Materials
900	Test Performances of Air Filter Units
340	Tests for Comparative Flammability of Liquids
72	Tests for Fire Resistance of Record Protection Equipment
790	Tests for Fire Resistance of Roof Covering Materials (DoD)
155	Tests for Fire Resistance of Vault and File Room Doors
214	Tests for Flame-Propagation of Fabrics & Films
94	Tests for Flammability of Plastic Materials for Parts in Devices & Appliances
991	Tests for Safety Related Controls Employing Solid State Devices
1439	Tests for Sharpness of Edges on Equipment

580	Tests for Uplift Resistance of Roof Assemblies [DoD]
1020	Thermal Cutoffs for use in Electrical Appliances and Components
547	Thermal Protectors for Motors
83	Thermoplastic-Insulated Wires and Cables [DoD]
863	Time Indicating and Recording Appliances
10A	Tin-Clad Fire Doors
697	Toy Transformers
1574	Track Lighting Systems
551	Transformer-Type Arc-welding Machines
1411	Transformers and Motor Transformers for use in Audio,Radio, and Television Type Appliances
1562	Transformers, Distribution, Dry-Type - Over 600 volts
1449	Transient Voltage Surge Suppressers [PE]
109	Tube Fittings for Flammable & Combustible Fluids, Refrigeration Svs, & Marine Use
651A	Type EB and A Rigid PVC Conduit and HDPE Conduit [DoD]
884	Underfloor Raceways and Fittings [DoD]
676	Underwater Lighting Fixtures
1778	Uninterruptible Power Supply Equipment
1062	Unit Substations
18	Unlined Fire Hose
647	Unvented Kerosene-Fired Room Heaters and Portable Heaters
1897	Uplift Tests for Roof Covering Systems
1017	Vacuum Cleaners and Blower Cleaners
125	Valves for Anhydrous Ammonia & LP-Gas (other Than Safety Relief)
842	Valves for Flammable Fluids
751	Vending Machines
17	Vent or Chimney Connector Dampers for Oil-Fired Appliances
1738	Venting Systems for Gas Burning Appliances, Categories II, III, and IV
1685	Vertical-Tray Fire Propagation and Smoke Release Test fo Electrical and Optical Fiber Cables
1638	Visual Signaling Appliances
430	Waste Disposers
296A	Waste Oil-Burning Air-Heating Appliances
346	Waterflow Indicators for Fire Protective Signaling Systems
1130	Wet Exhaust Components for Marine Engines
997	Wind Resistance of Prepared Roof Covering Materials [DoD]
486A	Wire Connectors & Soldering Lugs for Use with Copper Conductors [DoD]
486B	Wire Connectors for use with Aluminum Conductors
1565	Wire Positioning Devices
65	Wired Cabinets
870	Wireways, Auxiliary Gutters, & Associated Fittings
1681	Wiring Device Configurations
187	X-Ray Equipment

Appendix C

National Fire
Protection Association
(NFPA)

Publications in Numerical Order

1   Fire Protection Code

10   Portable Fire Extinguishers

10R   Portable Fire Extinguishers in Family Dwellings

11   Low Expansion Foam

11A   Medium & High Expansion Foam Systems

11C   Mobile Foam Apparatus

12A   Halon 1301 Fire Extinguishing Systems

13   Installation of Sprinkler Systems

13D   Sprinkler Systems in One & Two Family Dwellings and Manufactured Homes

13E   Fire Department Operations in Properties Protected by Sprinkler and Standpipe Systems

13R   Sprinkler Systems in Residential Occupancies up to and Including Four Stories in Height

14   Installation of Standpipe and Hose Systems

15   Water Spray Fixed Systems

16   Deluge Foam-Water Sprinkler Systems & Foam-Water Spray Systems

16A   Installation of Closed Head Foam-Water Sprinkler Systems

17   Dry Chemical Extinguishing Systems

17A   Wet Chemical Extinguishing Systems

18   Wetting Agents

20   Installation of Centrifugal Fire Pumps

22   Water Tanks for Private Fire Protection

24   Installation of Private Fire Service Mains and Their Appurtenances

25   Water-Based Fire Protection Systems

30   Flammable and Combustible Liquid Code

30A   Automotive and Marine Service Station Code

30B   Aerosol Products, Manufacture and Storage

31   Installation of Oil Burning Equipment

32   Drycleaning Plants

33   Spray Application Using Flammable or Combustible Materials

34	Dipping & Coating Processes Using Flammable or Combustible Liquids
35	Manufacture of Organic Coatings
36	Solvent Extraction Plants
37	Stationary Combustion Engines & Gas Turbines
40	Storage & Handling of Cellulose Nitrate Motion Picture Film
40E	Storage of Pyroxylin Plastic
43B	Organic Peroxide Formulations, Storage of
43D	Storage of Pesticides
45	Fire Protection for Laboratories Using Chemicals
46	Storage of Forest Products
49	Hazardous Chemicals Data
50	Bulk Oxygen Systems at Customer Sites
50A	Gaseous Hydrogen Systems and Customers Sites
50B	Liquefied Hydrogen Systems at Customer Sites
51	Design & Installation of Oxygen-Fuel Gas Systems for Welding, Cutting & Allied Processes
51A	Acetylene Cylinder Charging Plants
51B	Cutting and Welding Processes
52	Compressed Natural Gas (CNG) Vehicular Fuel Systems
53	Fire Hazards in Oxygen-Enriched Atmospheres
54	National Fuel Gas Code
55	Compressed & Liquefied Gases in Portable Cylinders
58	Storage & Handling of Liquefied Petroleum Gases
59	Storage & Handling of Liquefied Petroleum Gases at Utility Gas Plants
59A	Liquefied Natural Gas (LNG)
61	Fire & Dust Explosions in Agricultural & Food Products Facilities
65	Processing and Finishing of Aluminum
68	Venting of Deflagrations
69	Explosion Prevention Systems
70	National Electric Code
70B	Electrical Equipment Maintenance
70E	Electrical Safety Requirements for Employee Workplaces
72	National Fire Alarm Code
73	Residential Electrical Maintenance
75	Protection of Electronic Computer/Data Processing Equipment
77	Static Electricity
79	Electrical Standard for Industrial Machinery
80	Fire Doors & Fire Windows
80A	Exterior Fire Exposures
81	Fur Storage, Fumigation and Cleaning
82	Incinerators, Waste & Linen Handling Systems & Equipment

86   Ovens & Furnaces
86C   Industrial Furnaces Using a Special Processing Atmosphere
86D   Industrial Furnaces Using Vacuum as an Atmosphere
88A   Parking Structures
88B   Repair Garages
90A   Installation of Air Conditioning & Ventilating Systems
90B   Installation of Warm Air Heating & Air Conditioning Systems
91   Installation of Exhaust Systems for Air Conveying of Materials
92A   Smoke Controlled Systems
92B   Smoke Management Systems in Malls, Atria, Large Areas
96   Cooking Operations, Venting Control
97   Glossary of Terms Relating to Chimneys, Vents & Heating Producing Appliances
99   Health Facilities
99B   Hyprobaric Facilities
101   Life Safety Code
102   Grandstands, Folding & Telescopic Seating, Tents & Membrane Structures
105   Smoke-Control Door Assemblies
110   Emergency and Standby Power Systems
111   Stored Electrical Energy Emergency & Standby Power Systems
115   Laser Fire Protection
120   Coal Preparation Plants
121   Self-Propelled and Mobile Surface Mining Equipment
122   Fire Prevention and Control in Underground Metal and Nonmetal Mines
123   Fire Prevention and Control in Underground Bituminous Coal Mines
130   Fixed Guideway Transit Systems
150   Fire Safety in Racetrack Stables
170   Fire Safety Symbols
203   Roof Coverings and Roof Deck Constructions
204M   Smoke and Heat Venting
211   Chimneys, Fireplaces, Vents and Solid Fuel Burning Appliances
214   Water Cooling Tanks
220   Types of Building Construction
221   Fire Walls and Fire Barrier Walls
231   General Storage
231C   Rack Storage of Materials
231D   Storage of Rubber Tires
231E   Storage of Baled Cotton
231F   Storage of Roll Paper
232   Records, Protection of
232A   Archives and Records Centers
241   Construction, Alteration, and Demolition Operations

251   Fire Tests of Building Construction and Materials
252   Fire Tests of Door Assemblies
253   Test for Critical Radiant Flux of Floor Covering Systems Using a Radiant Heat Energy Source
255   Test of Surface Burning Characteristics of Building Materials
256   Methods of Fire Tests of Roof Coverings
257   Fire Tests of Window Assemblies
258   Research Test Method for Determining Smoke Generation of Solid Materials
259   Test Method for Potential Heat of Building Materials
260   Cigarette Ignition Resistance of Components of Upholstered Furniture
261   Method of Test for Determining Resistance of Mock-up Upholstered Furniture Material Assemblies in Ignition by Smoldering Cigarettes
262   Method of Test for Fire and Smoke Characteristics of Wires and Cables
263   Heat and Visible Smoke Release Rates for Materials and Products, Method for Test for
264   Heat and Visible Smoke Release Rates for Materials and Products using an Oxygen Consumption Calorimeter
264A Method of Test for Heat Release Rates for Upholstered Furniture Components cr Composites and Mattresses using an Oxygen Consumption Calorimeter
265   Evaluating Room Fire Growth Construction of Textile Wall Coverings
266   Fire Characteristics of Upholstered Furniture Exposed to Flaming Ignition Source, Method of Test for
267   Fire Characteristics of Mattresses and Bedding Assemblies Exposed to Flaming Ignition Source, Method of Test for
291   Fire Flow Testing and Marking of Hydrants
295   Wildfire Control
297   Principles and Practices for Communication Systems
298   Foam Chemicals for Class A Fuels - Rural, Suburban, Veg. Areas
299   Protection of Fife and Property from Wildfire
302   Pleasure and Commercial Motor Craft
303   Marines and Boatyards
306   Control of Gas Hazards on Vessels
307   Marine Terminals, Piers and Wharves
312   Fire Protection of Vessels During Construction, Repair, and Lay-ups
316   Protection of Cleanrooms
321   Basic Classification of Flammable and Combustible Liquids
325   Fire Hazard Properties of Flammable Liquids, Gases, and Volatile Solids
326   Safe Entry of Underground Storage Tanks
327   Cleaning or Safeguarding Small Tanks and Containers without Entry
328   Control of Flammable and Combustible Liquids and Gasses in Manholes, Sewers, and Similar Underground Structures

329 Handling Underground Releases of Flammable and Combustible Liquids
385 Tank Vehicles for Flammable and Combustible Liquids
386 Portable Shipping Tanks for Flammable and Combustible Liquids
395 Storage of Flammable and Combustible Liquids on Farms and Isolated Sites
402M Aircraft Rescue and Fire Fighting Operations
403 Aircraft Rescue and Fire Fighting Services and Airports
407 Aircraft Fuel Servicing
408 Aircraft Hand Portable Fire Extinguishers
409 Aircraft Hangers
410 Aircraft Maintenance
412 Evacuating Aircraft Rescue and Fire Fighting Foam Equipment
414 Aircraft Rescue and Fire Fighting Vehicles
415 Aircraft Fueling Ramp Drainage
416 Construction and Protection of Aircraft Terminal Buildings
417 Construction and Protection of Aircraft Loading Walkways
418 Heliports
419 Master Planning Airport Water Supply Systems for Fire Protection
422 Aircraft Accident Response
423 Construction and Protection of Aircraft Engine Test Facilities
424M Airport/Community Emergency Planning
430 Liquid and Solid Oxidizers
471 Responding to Hazardous Materials Incidents
472 Professional Competence of Responders to Hazardous Materials Incidents
473 Competencies for EMS Personnel
480 Storage, Handling and Processing of Magnesium
481 Production, Processing Handling and Storage of Titanium
482 Production, Processing, Handling and Storage of Zirconium
485 Lithium Metal - Storage, Handling, Processing, and Use
490 Storage of Ammonium Nitrate
491M Hazardous Chemical Reactions
495 Explosive Materials Code
496 Purged and Pressurized Enclosures for Electrical Equipment in Hazardous (Classified) Locations
497A Classification of Class I Hazardous Locations for Electrical Installations in Chemical Process Areas
497B Classification of Class II Hazardous (Classified) Locations for Electrical Installations in Chemical Processing Areas
497M Classification of Gasses, Vapors and Dusts for Electrical Equipment in Hazardous (Classified) Locations
498 Explosives Motor Vehicle Terminals
501A Fire Safety Criteria for Manufactured Home Installations, Sites and Communities

501C Firesafety Criteria for Recreational Vehicles
501D Firesafety Criteria for Recreational Vehicle Parks and Campgrounds
502   Fire Protection for Limited Access Highways, Tunnels, Bridges, Elevated Roadways, and Air Right Structures
505   Powered Industrial Trucks Including Type Designations, Areas of Use, Maintenance and Operations
512   Truck Fire Protection
513   Motor Freight Terminals
550   Fire Safety Concepts Tree
560   Ethylene Oxide for Sterilization and Fumigation, Storage, Handling, and Use of
600   Industrial Fire Brigades
601   Guard Service in Fire Loss Prevention
650   Pneumatic Conveying Systems for Handling Combustible Materials
651   Manufacture of Aluminum Power
654   Prevention of Fire and Dust Explosions in the Chemical Dye, Pharmaceutical, and Plastics Industries
655   Sulfur Fires and Explosions
664   Fires and Explosions in Wood Processing and Woodworking Facilities
701   Methods of Fire Tests for Flame-Resistant Textiles and Films
703   Fire Retardant Impregnated Wood and Fire Retardant Coating for Building Materials
704   Identification of the Fire Hazards of Materials
705   Field Flame Test for Textiles and Films
780   Installation of Lighting Protection Systems
801   Facilities Handling Radioactive Materials
802   Nuclear Research and Production Reactors
803   Light Water Nuclear Power Plants
804   Fire Protection for Advanced Light Water Reactor Electric Generating Plants
820   Fire Protection in Wastewater Treatment and Collection Facilities
850   Electric Generating Plants
851   Hydroelectric Generating Plants
901   Standard Classifications for Incident Reporting and Fire Protection Data
902M Fire Reporting Field Incident Manual
903   Fire Reporting Property Survey Guide
904   Incident Follow-up Report Guide
906   Fire Incident Field Notes
910   Libraries and Library Collections
911   Museums and Museum Collections
912   Places of Worship
913   Historic Structures and Sites
914   Fire Protection of Historic Structures
921   Fire and Explosion Investigations, Guide for

1000 Fire Service Professional Qualifications
1001 Fire Fighter Professional Qualifications
1002 Fire Department Vehicle Driver/Operator Professional Qualifications
1021 Airport Fire Fighter Professional Qualifications
1031 Fire Officer Professional Qualifications
1033 Fire Investigator Professional Qualifications
1035 Public Fire and Life Safety Educator Professional Qualifications
1041 Fire Service Instructor Professional Qualifications
1051 Wildland Fire Fighter Professional Qualifications
1122 Code for Modern Rocketry
1123 Fireworks Displays
1124 Manufacture, Transportation, and Storage of Fireworks
1125 Model Rocket and High Power Rocket Motors
1126 Use of Pyrotechnics Before a Proximate Audience
1127 High Power Rocketry
1141 Planned Building Groups
1201 Developing Fire Protection Services for the Public
1221 Installation, Maintenance, and Use of Public Fire Service Communication Systems
1231 Water Supplies for Suburban and Rural Fire Fighting
1401 Fire Protection Training Reports and Records
1402 Building Fire Service Training Centers
1403 Live Fire Training Evaluations in Structures
1404 Fire Department Self-Contained Breathing Apparatus Program
1405 Land-Basted Fire Fighters Who Respond to Marine Vessel Fires
1406 Outside Live Fire Training Evaluations
1410 Training for Initial Fire Attack
1420 Pre-Incident Planning for Warehouse Occupancies
1452 Training Fire Department Personnel to Make Dwelling Fire Safety Surveys
1470 Search and Rescue for Structural Collapse Incidents
1500 Fire Department Occupational Safety and Health Program
1521 Fire Department Safety Officer
1561 Fire Department Incident Management System
1581 Fire Department Infection Control Program
1582 Medical Requirements for Fire Fighters
1600 Disaster Management
1901 Pumper Fire Apparatus
1902 Initial Attack Fire Apparatus
1903 Mobile Water Supply Fire Apparatus
1904 Aerial Ladder and Elevating Platform Fire Apparatus
1906 Wildland Fire Apparatus
1911 Service Tests of Pumps on Fire Department Apparatus

1914   Fire Department Aerial Devices, Testing
1921   Fire Department Portable Pumping Units
1922   Fire Service Self-Contained Pumping Units
1931   Design of and Design Verification Tests for Fire Department Ground Ladders
1932   Use, Maintenance and Service Testing of Fire Department Ground Ladders
1961   Fire Hose
1962   Care, Use and Service Testing of Fire Hose Including Connections and Nozzles
1963   Fire Hose Connections
1964   Spray Nozzles (Shutoff and Tip)
1971   Protective Clothing for Structural Fire Fighting
1972   Helmets for Structural Fire Fighting
1973   Gloves for Structural Fire Fighting
1974   Protective Footwear for Structural Fire Fighting
1975   Station/Work Uniforms
1976   Protective Clothing for Proximity Fire Fighting
1977   Protective Clothing and Equipment for Wildland Fire Fighting
1981   Open-Circuit Self-Contained Breathing Apparatus for Fire Fighters
1982   Personal Alert Safety Systems (PASS) for Fire Fighters
1983   Fire Service Life Safety Rope and System Components
1991   Vapor-Protective Suits for Hazardous Chemical Emergencies
1992   Liquid Splash-Protective Suits for Hazardous Chemical Emergencies
1993   Support Function Protective Garments for Hazardous Chemical Operations
1999   Protective Clothing for Medical Emergency Operations
2001   Cleaning Agent for Fire Extinguishing Systems
8501   Single Burner Boiler Operations
8502   Prevention of Furnace Explosions/Implosions in Multiple Burner Boilers
8503   Pulverized Fuel Systems
8504   Atmospheric Fluidized-Bed Boiler Operations
8505   Stoker Operation
8506   Heat Recovery Steam Generator Systems

Appendix D

National Fire
Protection Association
(NFPA)

Publications in Alphabetical Order

51A   Acetylene Cylinder Charging Plants

1904   Aerial Ladder and Elevating Platform Fire Apparatus

30B   Aerosol Products, Manufacture and Storage

422   Aircraft Accident Response

407   Aircraft Fuel Servicing

415   Aircraft Fueling Ramp Drainage

408   Aircraft Hand Portable Fire Extinguishers

409   Aircraft Hangers

410   Aircraft Maintenance

402M Aircraft Rescue and Fire Fighting Operations

414   Aircraft Rescue and Fire Fighting Vehicles

403   Aircraft Rescue and Fire Fighting Services and Airports

1021   Airport Fire Fighter Professional Qualifications

424M Airport/Community Emergency Planning

232A Archives and Records Centers

8504 Atmospheric Fluidized-Bed Boiler Operations

30A   Automotive and Marine Service Station Code

321   Basic Classification of Flammable and Combustible Liquids

1402   Building Fire Service Training Centers

50   Bulk Oxygen Systems at Customer Sites

1962   Care, Use and Service Testing of Fire Hose Including Connections and Nozzles

211   Chimneys, Fireplaces, Vents and Solid Fuel Burning Appliances

260   Cigarette Ignition Resistance of Components of Upholstered Furniture

497A Classification of Class I Hazardous Locations for Electrical Installations in Chemical Process Areas

497B Classification of Class II Hazardous (Classified) Locations for Electrical Installations in Chemical Processing Areas

497M Classification of Gasses, Vapors and Dusts for Electrical Equipment in Hazardous (Classified) Locations

2001   Cleaning Agent for Fire Extinguishing Systems

327   Cleaning or Safeguarding Small Tanks and Containers without Entry

120	Coal Preparation Plants
1122	Code for Modern Rocketry
473	Compotencies for EMS Personnel
55	Compressed & Liquefied Gases in Portable Cylinders
52	Compressed Natural Gas (CNG) Vehicular Fuel Systems
423	Construction and Protection of Aircraft Engine Test Facilities
417	Construction and Protection of Aircraft Loading Walkways
416	Construction and Protection of Aircraft Terminal Buildings
241	Construction, Alteration, and Demolition Operations
328	Control of Flammable and Combustible Liquids and Gasses in Manholes, Sewers, and Similar Underground Structures
306	Control of Gas Hazards on Vessels
96	Cooking Operations, Venting Control
51B	Cutting and Welding Processes
16	Deluge Foam-Water Sprinkler Systems & Foam-Water Spray Systems
51	Design & Installation of Oxygen-Fuel Gas Systems for Welding, Cutting & Allied Processes
1931	Design of and Design Verification Tests for Fire Department Ground Ladders
1201	Developing Fire Protection Services for the Public
34	Dipping & Coating Processes Using Flammable or Combustible Liquids
1600	Disaster Management
17	Dry Chemical Extinguishing Systems
32	Drycleaning Plants
850	Electric Generating Plants
70B	Electrical Equipment Maintenance
70E	Electrical Safety Requirements for Employee Workplaces
79	Electrical Standard for Industrial Machinery
110	Emergency and Standby Power Systems
560	Ethylene Oxide for Sterilization and Fumigation, Storage, Handling, and Use of
412	Evacuating Aircraft Rescue and Fire Fighting Foam Equipment
265	Evaluating Room Fire Growth Construction of Textile Wall Coverings
80A	Exterior Fire Exposures
69	Explosion Prevention Systems
495	Explosive Materials Code
498	Explosives Motor Vehicle Terminals
801	Facilities Handling Radioactive Materials
705	Field Flame Test for Textiles and Films
61	Fire & Dust Explosions in Agricultural & Food Products Facilities
921	Fire and Explosion Investigations, Guide for
267	Fire Characteristics of Mattresses and Bedding Assemblies Exposed to Flaming Ignition Source, Method of Test for

266 Fire Characteristics of Upholstered Furniture Exposed to Flaming Ignition Source, Method of Test for
1914 Fire Department Aerial Devices, Testing
1561 Fire Department Incident Management System
1581 Fire Department Infection Control Program
1500 Fire Department Occupational Safety and Health Program
1921 Fire Department Portable Pumping Units
1521 Fire Department Safety Officer
1404 Fire Department Self-Contained Breathing Apparatus Program
1002 Fire Department Vehicle Driver/Operator Professional Qualifications
13E Fire Department Operations in Properties Protected by Sprinkler and Standpipe Systems
80 Fire Doors & Fire Windows
1001 Fire Fighter Professional Qualifications
291 Fire Flow Testing and Marking of Hydrants
325 Fire Hazard Properties of Flammable Liquids, Gases, and Volatile Solids
53 Fire Hazards in Oxygen-Enriched Atmospheres
1961 Fire Hose
1963 Fire Hose Connections
906 Fire Incident Field Notes
1033 Fire Investigator Professional Qualifications
1031 Fire Officer Professional Qualifications
123 Fire Prevention and Control in Underground Bituminous Coal Mines
122 Fire Prevention and Control in Underground Metal and Nonmetal Mines
1 Fire Protection Code
45 Fire Protection for Laboratories Using Chemicals
502 Fire Protection for Limited Access Highways, Tunnels, Bridges, Elevated Roadways, and Air Right Structures
820 Fire Protection in Wastewater Treatment and Collection Facilities
914 Fire Protection of Historic Structures
312 Fire Protection of Vessels During Construction, Repair, and Lay-ups
1401 Fire Protection Training Reports and Records
804 Fire Protection for Advanced Light Water Reactor Electric Generating Plants
902M Fire Reporting Field Incident Manual
903 Fire Reporting Property Survey Guide
703 Fire Retardant Impregnated Wood and Fire Retardant Coating for Building Materials
550 Fire Safety Concepts Tree
501A Fire Safety Criteria for Manufactured Home Installations, Sites and Communities
150 Fire Safety in Racetrack Stables
170 Fire Safety Symbols
1041 Fire Service Instructor Professional Qualifications

1983	Fire Service Life Safety Rope and System Components
1000	Fire Service Professional Qualifications
1922	Fire Service Self-Contained Pumping Units
251	Fire Tests of Building Construction and Materials
252	Fire Tests of Door Assemblies
257	Fire Tests of Window Assemblies
221	Fire Walls and Fire Barrier Walls
664	Fires and Explosions in Wood Processing and Woodworking Facilities
501D	Firesafety Criteria for Recreational Vehicle Parks and Campgrounds
501C	Firesafety Criteria for Recreational Vehicles
1123	Fireworks Displays
130	Fixed Guideway Transit Systems
30	Flammable and Combustible Liquid Code
298	Foam Chemicals for Class A Fuels - Rural, Suburban, Veg. Areas
512	Truck Fire Protection
81	Fur Storage, Fumigation and Cleaning
50A	Gaseous Hydrogen Systems and Customers Sites
231	General Storage
97	Glossary of Terms Relating to Chimneys, Vents & Heating Producing Appliances
1973	Gloves for Structural Fire Fighting
102	Grandstands, Folding & Telescopic Seating, Tents & Membrane Structures
601	Guard Service in Fire Loss Prevention
12A	Halon 1301 Fire Extinguishing Systems
329	Handling Underground Releases of Flammable and Combustible Liquids
491M	Hazardous Chemical Reactions
49	Hazardous Chemicals Data
99	Health Facilities
264	Heat and Visible Smoke Release Rates for Materials and Products using an Oxygen Consumption Calorimeter
263	Heat and Visible Smoke Release Rates for Materials and Products, Method for Test for
8506	Heat Recovery Steam Generator Systems
418	Heliports
1972	Helmets for Structural Fire Fighting
1127	High Power Rocketry
913	Historic Structures and Sites
851	Hydroelectric Generating Plants
99B	Hyprobaric Facilities
704	Identification of the Fire Hazards of Materials
904	Incident Follow-up Report Guide
82	Incinerators, Waste & Linen Handling Systems & Equipment

600	Industrial Fire Brigades
86C	Industrial Furnaces Using a Special Processing Atmosphere
86D	Industrial Furnaces Using Vacuum as an Atmosphere
1902	Initial Attack Fire Apparatus
90A	Installation of Air Conditioning & Ventilating Systems
20	Installation of Centrifugal Fire Pumps
16A	Installation of Closed Head Foam-Water Sprinkler Systems
91	Installation of Exhaust Systems for Air Conveying of Materials
780	Installation of Lighting Protection Systems
31	Installation of Oil Burning Equipment
24	Installation of Private Fire Service Mains and Their Appurtenances
13	Installation of Sprinkler Systems
14	Installation of Standpipe and Hose Systems
90B	Installation of Warm Air Heating & Air Conditioning Systems
1221	Installation, Maintenance, and Use of Public Fire Service Communication Systems
1405	Land-Basted Fire Fighters Who Respond to Marine Vessel Fires
115	Laser Fire Protection
910	Libraries and Library Collections
101	Life Safety Code
803	Light Water Nuclear Power Plants
50B	Liquefied Hydrogen Systems at Customer Sites
59A	Liquefied Natural Gas (LNG)
430	Liquid and Solid Oxidizers
1992	Liquid Splash-Protective Suits for Hazardous Chemical Emergencies
485	Lithium Metal - Storage, Handling, Processing, and Use
1403	Live Fire Training Evaluations in Structures
11	Low Expansion Foam
651	Manufacture of Aluminum Power
35	Manufacture of Organic Coatings
1124	Manufacture, Transportation, and Storage of Fireworks
307	Marine Terminals, Piers and Wharves
303	Marines and Boatyards
419	Master Planning Airport Water Supply Systems for Fire Protection
1582	Medical Requirements for Fire Fighters
11A	Medium & High Expansion Foam Systems
261	Method of Test for Determining Resistance of Mock-up Upholstered Furniture Material Assemblies in Ignition by Smoldering Cigarettes
262	Method of Test for Fire and Smoke Characteristics of Wires and Cables
264A	Method of Test for Heat Release Rates for Upholstered Furniture Components or Composites and Mattresses using an Oxygen Consumption Calorimeter
701	Methods of Fire Tests for Flame-Resistant Textiles and Films

256	Methods of Fire Tests of Roof Coverings
11C	Mobile Foam Apparatus
1903	Mobile Water Supply Fire Apparatus
1125	Model Rocket and High Power Rocket Motors
513	Motor Freight Terminals
911	Museums and Museum Collections
70	National Electric Code
72	National Fire Alarm Code
54	National Fuel Gas Code
802	Nuclear Research and Production Reactors
1981	Open-Circuit Self-Contained Breathing Apparatus for Fire Fighters
43B	Organic Peroxide Formulations, Storage of
1406	Outside Live Fire Training Evaluations
86	Ovens & Furnaces
88A	Parking Structures
1982	Personal Alert Safety Systems (PASS) for Fire Fighters
912	Places of Worship
1141	Planned Building Groups
302	Pleasure and Commercial Motor Craft
650	Pneumatic Conveying Systems for Handling Combustible Materials
10	Portable Fire Extinguishers
10R	Portable Fire Extinguishers in Family Dwellings
386	Portable Shipping Tanks for Flammable and Combustible Liquids
1974	Protective Footwear for Structural Fire Fighting
505	Powered Industrial Trucks Including Type Designations, Areas of Use, Maintenance and Operations
1420	Pre-Incident Planning for Warehouse Occupancies
654	Prevention of Fire and Dust Explosions in the Chemical Dye, Pharmaceutical, and Plastics Industries
8502	Prevention of Furnace Explosions/Implosions in Multiple Burner Boilers
297	Principles and Practices for Communication Systems
65	Processing and Finishing of Aluminum
481	Production, Processing Handling and Storage of Titanium
482	Production, Processing, Handling and Storage of Zirconium
472	Professional Competence of Responders to Hazardous Materials Incidents
316	Protection of Cleanrooms
75	Protection of Electronic Computer/Data Processing Equipment
299	Protection of Fife and Property from Wildfire
1977	Protective Clothing and Equipment for Wildland Fire Fighting
1999	Protective Clothing for Medical Emergency Operations
1976	Protective Clothing for Proximity Fire Fighting

1971	Protective Clothing for Structural Fire Fighting
1035	Public Fire and Life Safety Educator Professional Qualifications
8503	Pulvertzed Fuel Systems
1901	Pumper Fire Apparatus
496	Purged and Pressurized Enclosures for Electrical Equipment in Hazardous (Classified) Locations
231C	Rack Storage of Materials
232	Records, Protection of
88B	Repair Garages
258	Research Test Method for Determining Smoke Generation of Solid Materials
73	Residential Electrical Maintenance
471	Responding to Hazardous Materials Incidents
203	Roof Coverings and Roof Deck Constructions
326	Safe Entry of Underground Storage Tanks
1470	Search and Rescue for Structural Collapse Incidents
121	Self-Propelled and Mobile Surface Mining Equipment
1911	Service Tests of Pumps on Fire Department Apparatus
8501	Single Burner Boiler Operations
204M	Smoke and Heat Venting
92A	Smoke Controlled Systems
92B	Smoke Management Systems in Malls, Atria, Large Areas
105	Smoke-Control Door Assemblies
36	Solvent Extraction Plants
33	Spray Application Using Flammable or Combustible Materials
1964	Spray Nozzles (Shutoff and Tip)
13D	Sprinkler Systems in One & Two Family Dwellings and Manufactured Homes
13R	Sprinkler Systems in Residential Occupancies up to and Including Four Stories in Height
901	Standard Classifications for Incident Reporting and Fire Protection Data
77	Static Electricity
1975	Station/Work Uniforms
37	Stationary Combustion Engines & Gas Turbines
8505	Stoker Operation
40	Storage & Handling of Cellulose Nitrate Motion Picture Film
58	Storage & Handling of Liquefied Petroleum Gases
59	Storage & Handling of Liquefied Petroleum Gases at Utility Gas Plants
490	Storage of Ammonium Nitrate
231E	Storage of Baled Cotton
395	Storage of Flammable and Combustible Liquids on Farms and Isolated Sites
46	Storage of Forest Products
43D	Storage of Pesticides

40E   Storage of Pyroxylin Plastic
231F   Storage of Roll Paper
231D   Storage of Rubber Tires
480   Storage, Handling and Processing of Magnesium
111   Stored Electrical Energy Emergency & Standby Power Systems
655   Sulfur Fires and Explosions
1993   Support Function Protective Garments for Hazardous Chemical Operations
385   Tank Vehicles for Flammable and Combustible Liquids
253   Test for Critical Radiant Flux of Floor Covering Systems Using a Radiant
       Heat Energy Source
259   Test Method for Potential Heat of Building Materials
255   Test of Surface Burning Characteristics of Building Materials
1452   Training Fire Department Personnel to Make Dwelling Fire Safety Surveys
1410   Training for Initial Fire Attack
220   Types of Building Construction
1126   Use of Pyrotechnics Before a Proximate Audience
1932   Use, Maintenance and Service Testing of Fire Department Ground Ladders
1991   Vapor-Protective Suits for Hazardous Chemical Emergencies
68   Venting of Deflagrations
214   Water Cooling Tanks
15   Water Spray Fixed Systems
1231   Water Supplies for Suburban and Rural Fire Fighting
22   Water Tanks for Private Fire Protection
25   Water-Based Fire Protection Systems
17A   Wet Chemical Extinguishing Systems
18   Wetting Agents
295   Wildfire Control
1906   Wildland Fire Apparatus
1051   Wildland Fire Fighter Professional Qualifications